Studies in International Performa

Published in association with the h

General Editors: **Janelle Reinelt** a

Culture and performance cross t lers
that define nations. In this new series, scholars of performance produce interactions between and among nations and cultures as well as genres, identities and
imaginations.

Inter-national in the largest sense, the books collected in the *Studies in International
Performance* series display a range of historical, theoretical and critical approaches
to the panoply of performances that make up the global surround. The series
embraces 'Culture' which is institutional as well as improvised, underground or
alternate, and treats 'Performance' as either intercultural or transnational as well
as intracultural within nations.

Titles include:

Patrick Anderson and Jisha Menon (*editors*)
VIOLENCE PERFORMED
Local Roots and Global Routes of Conflict

Elaine Aston and Sue-Ellen Case
STAGING INTERNATIONAL FEMINISMS

Christopher Balme
PACIFIC PERFORMANCES
Theatricality and Cross-Cultural Encounter in the South Seas

Matthew Isaac Cohen
PERFORMING OTHERNESS
Java and Bali on International Stages, 1905–1952

Susan Leigh Foster (*editor*)
WORLDING DANCE

Helen Gilbert and Jacqueline Lo
PERFORMANCE AND COSMOPOLITICS
Cross-Cultural Transactions in Australasia

Helena Grehan
PERFORMANCE, ETHICS AND SPECTATORSHIP IN A GLOBAL AGE

Judith Hamera
DANCING COMMUNITIES
Performance, Difference, and Connection in the Global City

James Harding and Cindy Rosenthal (*editors*)
THE RISE OF PERFORMANCE STUDIES
Rethinking Richard Schecher's Broad Spectrum

Silvija Jestrovic and Yana Meerzon (*editors*)
PERFORMANCE, EXILE AND 'AMERICA'

Ola Johansson
COMMUNITY THEATRE AND AIDS

Ketu Katrak
CONTEMPORARY INDIAN DANCE
New Creative Choreography in India and the Diaspora

Sonja Arsham Kuftinec
THEATRE, FACILITATION, AND NATION FORMATION IN THE BALKANS
AND MIDDLE EAST

Daphne P. Lei
ALTERNATIVE CHINESE OPERA IN THE AGE OF GLOBALIZATION
Performing Zero

Carol Martin (*editor*)
THE DRAMATURGY OF THE REAL ON THE WORLD STAGE

Alan Read
THEATRE, INTIMACY & ENGAGEMENT
The Last Human Venue

Shannon Steen
RACIAL GEOMETRIES OF THE BLACK ATLANTIC, ASIAN PACIFIC AND
AMERICAN THEATRE

Joanne Tompkins
UNSETTLING SPACE
Contestations in Contemporary Australian Theatre

S. E. Wilmer
NATIONAL THEATRES IN A CHANGING EUROPE

Evan Darwin Winet
INDONESIAN POSTCOLONIAL THEATRE
Spectral Genealogies and Absent Faces

Forthcoming titles:

Adrian Kear
THEATRE AND EVENT

Studies in International Performance
Series Standing Order ISBN 978–1–4039–4456–6 (hardback)
978–1–4039–4457–3 (paperback)
(*outside North America only*)

You can receive future titles in this series as they are published by placing a standing order. Please contact your bookseller or, in case of difficulty, write to us at the address below with your name and address, the title of the series and the ISBN quoted above.

Customer Services Department, Macmillan Distribution Ltd, Houndmills, Basingstoke, Hampshire RG21 6XS, England

---

Dancing Communities

Performance, Difference and Connection in the Global City

Judith Hamera

First published 2007
Published in paperback 2011 by
PALGRAVE MACMILLAN

Palgrave Macmillan in the UK is an imprint of Macmillan Publishers Limited, registered in England, company number 785998, of Houndmills, Basingstoke, Hampshire RG21 6XS.

Palgrave Macmillan in the US is a division of St Martin's Press LLC, 175 Fifth Avenue, New York, NY 10010.

Palgrave Macmillan is the global academic imprint of the above companies and has companies and representatives throughout the world.

Palgrave® and Macmillan® are registered trademarks in the United States, the United Kingdom, Europe and other countries.

ISBN 978-0-230-00003-2 hardback
ISBN 978-0-230-30233-4 paperback

This book is printed on paper suitable for recycling and made from fully managed and sustained forest sources. Logging, pulping and manufacturing processes are expected to conform to the environmental regulations of the country of origin.

A catalogue record for this book is available from the British Library.

Library of Congress Cataloging-in-Publication Data
Hamera, Judith.
   Dancing communities : performance, difference, and connection in the global city / Judith Hamera.
       p. cm.
   ISBN 978-0-230-00003-2 (cloth) 978-0-230-30233-4 (pbk)
   1. Dance—Social aspects.   2. Dance—Psychological aspects.
   3. Performing arts—Social aspects.   4. Communication—Social aspects.   I. Title.
   GV1595.H32 2007
   306.4'84—dc22   2006047263

10   9   8   7   6   5   4   3   2   1
20   19   18   17   16   15   14   13   12   11

Printed and bound in Great Britain by
CPI Antony Rowe, Chippenham and Eastbourne

For Alfred, best partner ever

'As a practitioner of several disciplines... and as a resident of the global city wherein Hamera grounds her investigation, I recommend Dancing Communities not only as a text worth reading but, even more so, as a task worth rehearsing.'

– Shakina Nayfack, Dance Research Journal

Contents

List of Illustrations

Series Preface

The 'Studies in International Performance' series was initiated in 2004 on behalf of the International Federation for Theatre Research, by Janelle Reinelt and Brian Singleton, successive Presidents of the Federation. Their aim was, and still is, to call on performance scholars to expand their disciplinary horizons to include the comparative study of performances across national, cultural, social, and political borders. This is necessary not only in order to avoid the homogenizing tendency of national paradigms in performance scholarship, but also in order to engage in creating new performance scholarship that takes account of and embraces the complexities of transnational cultural production, the new media, and the economic and social consequences of increasingly international forms of artistic expression. Comparative studies (especially when conceived across more than two terms) can value both the specifically local and the broadly conceived global forms of performance practices, histories, and social formations. Comparative aesthetics can challenge the limitations of national orthodoxies of art criticism and current artistic knowledges. In formalizing the work of the Federation's members through rigorous and innovative scholarship this series aims to make a significant contribution to an ever-changing project of knowledge creation.

Janelle Reinelt and Brian Singleton
International Federation for Theatre Research
Fédération Internationale pour la Recherche
Théâtrale

Preface to the Paperback Edition

Dancing Communities makes two main points. First, dance is vital urban infrastructure. It produces generative and unexpected alliances across multiple dimensions of difference and creates vernacular spaces to both complicate and humanize the transnational concentrations and flows characterizing the global city. Second, technique makes dance go. Whether it is embraced or resisted, master or mastered, it is the discursive currency that sustains amateur and professional communities. Further, dance techniques, however codified or 'elite,' are routine tactics for living, not simply abstract grammars for moving. As practices of everyday community life, they build connections between what dancers do onstage and off, how and what they feel about their work and their lives, and ways they reflect on their relationships to their bodies, to history, to the ineffable, and to one another.

The five communities examined here are concrete examples of how performance as a world-making endeavor succeeds and sometimes fails. I believe the lessons these artists teach us are even more consequential and urgent now than when the first edition of this book was published. Two contemporary events are illustrative.

In 2008, the *Los Angeles Times* eliminated its full-time dance critic position in one of a seemingly endless series of staff contractions. Though the paper still publishes dance reviews, the move significantly diluted a stable, high profile catalyst for mainstream analyses of L.A. dance, limiting debates about how to assess it and insert it into personal, local, and global aesthetic histories and traditions. In a city famous for its own amnesia, it eliminated a consistent and reliable archival voice and, to the extent that the paper circulated beyond the L.A. basin, it reduced local dancers' national and international profiles by, among other things, constricting a prominent structural platform for advocacy and visibility. With this action, the *Times* effectively wrote dance out of its institutional commitment to cover and examine key civic discourses.

At the same time, dance was – and still is – proliferating on network television, increasing exponentially in quantity if not variety. Movement vocabularies are similar, in their broad contours, across these dance shows: obviously athletic and spectacular, heteronormative, often strongly narrative, and highly emotional. The shows' internal critical responses include enthusiastic cheers from studio audiences to serve as paralinguistic highlighting of clearly affecting moments or equally clearly virtuosic execution, and near universal ascription of dancers' abilities to individual ethos in formulations like the seemingly self-evident 'living the dance,' or 'talent,' or self-expressive impulse. Above all, these shows share the frame of competition and with it the deleterious fiction that achieving excellence in performance is a zero sum game with clear winners and losers who succeed or fail despite or because of obvious merit, obvious pandering to one constituency or another, or obvious caprice.

My point in pairing these two recent phenomena is not to mourn the loss of an outlet for high art critique while bemoaning the banality of its popular relative but to examine where they converge: at the intimate coupling of dance and the precarious. Of course dance and precariousness have never been strangers, as anyone who has ever struggled to hold balance can attest. Indeed, repeated negotiations with the precarious animate dance and dancers whether in struggles against corporeal limits and aging, patching together a decent living from one's creative work, or managing physically or psychically risky moves as a soloist or ensemble member.

The precariousness I mean, while including all of these elements, is larger: part and parcel of our current global political economic context. It is, in a way, the bookend to the 'all-that-is-solid-melts-into-air' moments marking the painful births of urban industrial modernity. As with its predecessor, the social force of today's precariousness is especially concentrated in large, diverse urban centers. This is the profound dislocation that comes with continual reminders of one's utter dispensability as artist, citizen, employee, and member of a public – indeed the dispensability of the very idea of a public outside of market segmentation – coupled with the economic imperative to incarnate and produce excellence all the time or else. Today's precariousness makes it grossly explicit across multiple channels of public rhetoric that contemporary social life itself is a zero sum game with clear winners and losers who succeed or fail despite or because of obvious merit, obvious pandering, or obvious caprice. The putative meritocracy of televised dance shows reminds us that, in the most corrosive sense, we have only our own 'talents' to raise as fragile bulwarks against our own dispensability, while 'winning' means donning the red shoes of neoliberal economic logics and dancing till we drop.

The consequences of these logics are as unevenly distributed as the rewards, and Los Angeles is an especially compelling city in which to map them. Like its western global counterparts, over the last 25 years of the 20th century L.A. traded factories for finance in a gradual de-emphasis on domestic industrial policy, particularly regarding manu-facturing, with the concomitant demonization of labor unions and driving down of domestic wages. Increasing plant closures, and partic-ularly the collapse of the aerospace industry, led to area job losses in the hundreds of thousands during part of the period covered by this book, especially 1990–95. For much of the early 1990s, the region exper-ienced near double digit unemployment even as it became increasingly demographically diverse and stratified. While the late 1990s and early 2000s saw some rebound, albeit highly unequally distributed, the Great Recession beginning in 2007 added devastating stresses to the region's already precarious political economic tectonics. 'Austerity' has emerged as the consensus political response, and implementation has focused on further contracting the public sector.

Many of the artists in this book have been directly affected by these convulsions. For example, the Sems (pseudonym), discussed in Chapter 3, were haunted by more than the ghosts of their classical dance teachers and memories of their Khmer Rouge torturers. They were also stalked relentlessly by lack: lack of employment opportun-ities for semi-skilled workers that paid wages sufficient to even support the cost of commuting to the job, lack of child care and after-school programs, lack of accessible and affordable health care – especially for acute mental distress, and lack of an adequate and coherent social safety net. The Sems were, for all intents and purposes, wholly on their own and bereft of social support: unable to muster the psychic strength and material resources necessary to navigate a dizzying patchwork of private, community and government agencies all, no doubt, well-intentioned.

And that patchwork, such as it was in the early 1990s, looks almost luxurious 20 years later, with only a few frayed threads still in place. Now, lack threatens a much broader demographic than the very poor and the very damaged. Hae Kyung Lee's (Chapter 4) California State University, Los Angeles students who entered the campus in 2007 when the first edition of this book was published will have endured an over 70 percent increase in tuition and fees by spring 2011 on top of faculty furloughs and reduced admissions and course offerings. For those in the university's economically depressed service area, these fee increases and service cuts structurally undermine one of the last remaining bridges to a middle-class wage. In this diminished context, auditioning for a dance-based talent show *cum* lottery could actually be a more viable, and certainly only slightly less precarious investment in one's future than counting on the state's Master Plan and its now empty affirmation of tuition-free higher education for all qualified residents.

Of course, as the excision of the *L.A. Times* dance critic position attests, the arts, and particularly non-elite forms, have been continually reminded of their own precariousness within the public sphere when they have not been dismissed outright as wholly marginal. For example, the Chairman of the US National Endowment for the Arts at this writing recently commented publicly that there were, in effect, too many theatres for the 'demand' and the number should be reduced. That a varied and robust theatre culture might be a crucial component of civic infrastructure, and that his own organization's mission might be construed as educating the public to this fact thereby increasing 'demand,' seems to have eluded him despite this being his job.

Dance in the global city, both as a concert form and particularly as amateur and professional practices of everyday life, offers valuable working alternatives to these ideologies of easy dispensability, to what Paolo Virno has aptly termed oppressive 'servile virtuosity,' and to a disembodied, evacuated public sphere. While the specific cases discussed in the pages that follow provide varied and concrete examples, one overarching point merits reinforcement.

Dance explicitly links aesthetics, affect, labor, and bodies together and insists on their centrality to a meaningful public discourse that reflects on our shared humanity and imagines alternative social alliances and arrangements. That particular artists or pieces may be more or less resistant to hegemonic forces is second to the reality that dancing communities offer potent challenges to rhetorics asserting that individual precariousness is an immutable structural fact best addressed by *laissez-faire* market logics, and that a rich, diverse public sphere is an unsupportable burden. Unexamined textual primacy, latent antitheatrical prejudice, mystification of relationships between corporeality and language, and pervasive bias toward economic elites have so constricted the conventional view of public discourse that the simple fact of dance as a rhetorical actor crucial to robust civic life is not patently obvious.

Consider just a few of the most challenging political and ethical questions concert dance alone has examined over the past 50 years: What are the limits and possibilities of race and gender mutability on the public stage, broadly construed? What do various regimes of work look like and feel like, and what are the consequences of particular strategies for organizing labor? How do illness and physical limits intervene in conventional aesthetics and politics of representation, mindful of Stuart Hall's admonition that how people are represented is often how they are treated? What are our obligations to the bodies that came before us and to those dancing alongside us? Many of the artists discussed in this book have made examinations of these questions the foci of their careers, as have the critics who write about them. These artists and critics do not make dance rhetorical and consequential. They recognize that it is so, then enlarge the reach and resonance of a particular performance by inserting it into broader aesthetic, historical and social matrices in ways that exactly parallel the standard financial and political analyses central to informed social life.

But of course rhetorical efficacy and force, and social import, are not limited to concert dance. The slow, accretional build-up of everyday time and talk that sustains all dance practice is itself potent and consequential: offering models for working together and for ministering to large and small dislocations and losses. The questions dancers examine in their daily rituals and conversations also resonate broadly, with implications for how we might think about ourselves against this seemingly all-pervasive contemporary precariousness. How is my body, in all its particularity, an analytic for exploring and defining my life? What is partnership and what is obligation? How can I converse with historical precedent? Is loss simple disappearance, or something else? When is hope a radical, transformative force, righteous anger a pious duty, and resignation an understandable refuge?

At their best, dancing communities offer multiple points of entry into the creative coalitions that literally beget social movements. Even the most utopian of them embrace the myriad messy vexations of time, place, and others' complex humanity. Even the most fractured assert the redemptive possibilities of public bodies answering back to loss. All the communities in this book materially enact three commitments on stage and off, albeit some more successfully than others: Daily touch, talk, desires, and dreams link private obligations for one another to public enactments. Shared fragilities are inseparable from shared strengths. And even the most virtuosic solos are birthed by mutuality, including the unsteady mutuality of spectatorship. Especially in precarious times, we must insist that these commitments are as indispensable to the very highest levels of public deliberation as they are to community theaters and storefront studios.

* * *

My dear friend and colleague D. Soyini Madison has been a generous and forceful advocate for this book since it was a seed of an idea for a conference paper. This edition is for her, with thanks.

Acknowledgements

Any ethnographic project that spans 15 years inevitably accumulates debts. All of mine are joyful.

I am profoundly grateful to the artists whose words, works and lives inspired this book and are discussed in these pages. In addition to providing me with what Kenneth Burke termed 'equipment for living,' their aesthetic and personal integrity was sustaining in every way. I only hope that this modest turn in our lengthy conversations in interviews, in class, and in spectatorship, is up to those they so kindly offered me.

Roxanne Steinberg and Melinda Ring offered the first dance class I took in Los Angeles. I am very grateful for their generosity and their exacting introduction to Body Weather Laboratory, and for our conversations over the years. As Chapter 1 indicates, Roxanne Steinberg and Oguri are always inspiring. They not only rewrite possibilities for movement for me, but those for rigor and gentle precision in art and life.

Charles and Philip Fuller are dear to me beyond measure. They are extraordinary artists, businessmen and friends, and they are much more. At Le Studio, they are architects and sustaining spirits of an extraordinarily nurturing sociality. Their work in support of this sociality is truly never done. Everything they do, from teaching class, to directing productions, to driving out to isolated freeways in the middle of the night to fill gas tanks, to visiting senior citizen students in jail are all completed with intelligence, modesty, open-heartedness and good humor. I am so grateful to know them. I also want to thank Gilma Bustillo, Jeanne King, Xavier Roncalli and Bill Sato, wonderful interlocutors about 'the great social experiment,' that Xavier calls Le Studio.

The Sem family, especially May and her daughters, were open to my intrusions beyond anything one could reasonably expect. I am very grateful for this and wish them all the happiness and peace possible in a troubled world. May they find Apsaras wherever and whenever they seek them.

It is impossible to overstate Hae Kyung Lee's importance to this project over the years. From the luminous beauty of her dances to the rigor of her spirit, she offers the absolute model of a life of grace and/in hard work, spirit in/as practice, art in/as life. She transcends every binary opposition any critic has ever used to talk about dance, or art or mindful living. I cherish our friendship.

I have also been blessed with brilliant, energetic, and generous performance and cultural studies colleagues who have consistently provided support and valuable advice at every stage of this project. Thanks to Janelle Reinelt and Brian Singleton, editors of Palgrave's *Studies in International Performance*, for their belief in this project and for their insightful feedback. Thanks, too, to the editorial and production staff at Palgrave Macmillan, especially Penny Simmons and Christabel Scaife, for their thoughtful work on the manuscript.

Jill Dolan and Della Pollock provided rigorous, valuable, and heartening comments on sections of the manuscript. What a privilege to have such precise and lyrical interlocutors who engage performance both cognitively and soulfully.

I am very grateful to Susan Mason for our conversations about performance, dance, and the arts in Los Angeles over many years. Her political compass, and her solid critic's eye, are most appreciated. Thanks too to Lisa Merrill for support and bracing conversations on performance, gender and history. Stephanie Nelson kindly shared my tales from the field and always offered useful advice. As the works I have cited show, I am also indebted to Diane Sippl, whose publications and conversation consistently challenged and sharpened my thinking about art and performance. Gust Yep's critical discussions of race, culture and sexuality were very helpful in formulating my analyses of these dimensions of difference.

Thanks to colleagues whose works and conversation contributed to my thinking about performance and sociality: Harris Berger, Michael Bowman, Ruth Laurion Bowman, Donnalee Dox, Leonard C. Hawes, Andrea Imhoff, Shannon Jackson, E. Patrick Johnson, Sonja Kuftinec, Kristin Langellier, Tim Miller, Eric Peterson, Peggy Phelan, Craig Gingrich Philbrook, Carl Selkin, Stacy Wolf, and Kathryn Woodard. Special thanks to Stephen Balfour for technical assistance, cheerfully and masterfully given.

I am especially grateful to Bryant Alexander, dear friend and fellow ethnographer. Our conversations, his wisdom and generosity of spirit, and his thoroughly inexhaustible sense of humor sustained me through the most joyful and the most devastating periods of this research. In his life and in his work, he always reminds me of the intimate relationship between personal integrity and scholarly acumen.

I owe a great debt to D. Soyini Madison. Her advice on, and her unflagging belief in, this project from its earliest beginnings were heartening in the most profound ways. Her rigorous approaches to critical ethnography and globalization, and her abiding love of performance, are both inspiring and instructive. I learn so much from her, not only about these matters, but also about friendship, loyalty and the deep structures of collegiality.

Fifteen years is enough time to see a generation of artists, colleagues and students come, go, and even stay. Those who have stayed and continue to make their lives in dance and/or performance studies merit special mention. Their commitment to their work offers that most valuable intellectual and aesthetic commodity: hope for what comes next. Thanks to Erica de la O, Claudia Lopez and Miguel Olvera, extraordinary artists all, and inspiration for yet another new generation. Thanks to Kishisa Ross, Jose Reynoso, the late Jill Yip, Dustin Abraham, Kamran Afary, D. Robert DeChaine, Tony Fitzgerald, Mandy Gamble, Richie Hao, Javon Johnson, Mike Kalustian, Kenneth Lee, Tina McDermott, Heidi Miller, Josh Miller and Jeff Vowell for stimulating conversations on stage, on the page, or in class. Special thanks to Diana Fisher for our wonderful discussions of space and place in Los Angeles.

Friendships endure as surely as new generations of artists and scholars come, go, and stay. Two good friends merit special thanks. Lesley DiMare has offered wonderful conversation and generous hospitality through all stages of this project and beyond. She and Joan Cashion have been supportive in every way. After so many years of hearing about this project, I'm sure they are as glad as I am to have the book completed. Lee Roloff and the late Dwight Conquergood offered me complementary models of scholarly and personal integrity for which I am deeply grateful. As my teachers and my friends, these extraordinary men stand at the center of my intellectual life.

I'm not sure what my parents, Thaddeus and Dolores Hamera, expected when they enrolled me in Grace Thomas's Little Studio of Dance for ballet lessons at age 4. I am quite sure it was neither a PhD in Performance Studies, nor 15 years of watching and writing about dance in Los Angeles. Nevertheless, my mother dutifully drove me to class weekly, then daily; sewed costumes; and waited through endless hours of class, or auditions, or rehearsals. My father diligently paid for pointe shoes and fees, put up a barre in the basement, and filmed countless recitals. This was my earliest introduction to relationships between dance and labor, to the work of performance. Thus, in so many ways, my parents were the start of it all, and I owe them a debt I can never repay.

Alfred Bendixen took up new versions of these same labors. He attended hundreds of performances; read and heard every version of every chapter, probably dozens of times; and engaged the ethnographic spirit, and all the demands and detours it entails, with the same enthusiasm, intellectual depth and deep good humor he brings to everything he does. Simply put: without him, not.

*   *   *

Research for this book was supported in part by a Creative Leave and Difference-in-Pay Leave from California State University, Los Angeles. Sections of *Dancing Communities*, in all cases significantly expanded, rethought and revised, have appeared in the following journals: portions of Chapter 1 were published as 'I Dance to You: Reflections on Luce Irigaray's *I Love to You* in Pilates and Virtuosity,' *Cultural Studies*, 15.2 (April, 2001): 229–40, and 'The Romance of Monsters: Theorizing the Virtuoso Body in Performance,' *Theatre Topics*, 10.2 (Fall, 2000): 145–53. Portions of Chapter 2 were published as 'All the (Dis)comforts of Home: Place, Gendered Self-fashioning, and Solidarity in a Ballet Studio,' *Text and Performance Quarterly*, 25.2 (April, 2005): 93–112, and 'The Ambivalent, Knowing Male Body in the Pasadena Dance Theatre,' *Text and Performance Quarterly* (July, 1994): 197–209. Portions of Chapter 3 were published as: 'An Answerability of Memory: "Saving" Khmer Classical Dance,' *TDR: The Drama Review*, 46.4 (Winter, 2002, T176): 65–85, and 'Body, Memory and Wordless Stories,' *Women and Language* (Summer, 1996): 64–8. Part of Chapter 4 was published as 'Dancing Other Wise: Ethics, Metaphysics and Utopia in Hae Kyung Lee and Dancers,' *Modern Drama*, 47.2 (Summer, 2004): 290–308.

Introduction: Dancing the City

Increasingly, the exploration of modern urban life requires the routine deployment of what in the past would have been regarded as purely artistic modes of understanding...

(Amin and Thrift, *Cities*, 2002)

Every day, urban communities are danced into being. This is more than a metaphor. It is a testament to the power of performance as a social force, as cultural poesis, as communication infrastructure that makes identity, solidarity and memory sharable. Communities are danced into being in daily, routine labor, time and talk backstage and off – sometimes way off – stage, as well as onstage. They are danced into being by virtuoso technicians and earnest amateurs. Diverse, generative urban communities emerge at dance's busy intersections of discipline and dreams, repetition and innovation, competition and care.

Aristotle observed, 'Friendship would seem to hold cities together' (8.1155a 23). Dancers make cities as friends, as partners, as *corps* and, in so doing, remake themselves, their audiences, and each other every day, day after day. They create 'civic culture' in multiple senses of the term: as a commodity for public consumption and as a vernacular connecting individuals who may otherwise have little in common. In performance as a final product, and especially as a daily discipline, dancers reach across multiple dimensions of difference to incarnate new shared aesthetic and social possibilities.

Dancing Communities argues that both concert dance and amateur practice are laboratories for examining and revisioning the myriad complex interrelations between gender, sexuality, race, class, and culture in urban life. Both offer more than a simple 'politics of propinquity.'[1] On the contrary, the work of dance exposes aesthetic spaces and practices

1

as social and vernacular, as sites where participants actively confront and engage tradition, authority, corporeality, and irreducible difference. The resulting arrangements are processes, not things, hence the use of the gerund: *dancing* communities. They are constituted by doing dance: making it, seeing it, learning it, talking, writing and fantasizing about it.

Urban planners and performance scholars often see contemporary relationships between performance and urban life in economic and demographic terms.[2] For example, Susan Bennett writes:

> In thinking through 'theatre and the city,' we need to pay attention to the proliferation of 'beneficiary' commerce in areas with a recognized arts profile. How does a city's relationship to the arts contribute to its understanding of commercial development (and vice versa), especially in the areas of leisure-targeted businesses such as restaurants, bars, commercial art galleries, bookstores, specialty stores, and souvenir shops in the theatres themselves? . . . What audiences come to these city theatres and how do they identify with the theatre space?
> ('Comment', 2001: n.p.)

While there is much to be gained by this line of inquiry, it is limited by several assumptions: that significant urban art-making happens in or near public commercial venues, that its value lies in generating a particular kind of product closely tied to leisure and its attendant forms of consumption, and that audiences are the primary consumers and beneficiaries.

Though these assumptions are certainly reasonable, a problem emerges when they are taken for the sum total of how, where and for whom urban performance 'works.' The transformative potential of art making in the city becomes severely constrained. As Ash Amin and Nigel Thrift argue in *Cities*, this economic-demographic view of urban performance and urban life generally, emphasizes 'spaces of tolerance and sociability, perhaps gathering points on particular occasions. They are not the formative spaces of hybrid identities and politics' (2002: 137). Amin and Thrift argue that new social arrangements require 'meaningful and repeated contact, the slow experience of working, being, and living with others, and the everyday fusion of cultures in what we consume, what we see, where we travel, how we live, with whom we play, and so on' (2002: 137). This may mean deep or repeated engagements with a particular performer or ensemble onstage. It certainly includes the daily social, emotional and political work that makes dance go offstage.

Dancing Communities focuses on these engagements and this work. In so doing, it argues forcefully for recasting dance aesthetics, and aesthetics generally, as practices of everyday urban life.

Aesthetics, technique, and the social lives of dance

Aesthetics are inherently social. The formal properties and presumptions intrinsic to the production and consumption of art are communicative currency developed by and circulating between artists, audiences and critics, binding them together in interpretive communities, serving as bases for exchange in the public and private conversations that constitute art's relational, political and affective lives.

Aesthetics offer vocabularies for exploring how art works and how it generates meaning. But even more importantly, aesthetic principles, values and vocabularies organize *where* art works and produces meaning in social time and social space. Aesthetics are integral to finished creative products, but also to the myriad ancillary socialities that never take the stage. Scholars have argued forcefully for more nuanced, politicized readings of the relationships between art, society and culture (see Matthews and McWhirter, 2003). Central to all such readings are aesthetics, the animating principles of art's social lives, not only in the traces it leaves behind but in the routine transactions of those for whom art making is, and happens in, a neighborhood, a set of corporeal possibilities, comforts and constraints linking private self-fashioning and community practice.[3]

The social work of aesthetics is especially central to performance, where labors of creation are explicitly communal and corporeal, and where corporeality and sociality are remade as surely as a formal event is produced. Here the contingent, situated nature of art's norms and pleasures is exposed with special clarity. These norms and pleasures are literally incarnated, embraced or resisted by particular bodies in specific places and times. In such corporeal, contextual specificity, aesthetics can never be mistaken for transcendent or timeless. It is always already entangled 'in systems of power, repression, and exclusion' (Matthews and McWhirter, 2003: xxvi). It exposes questions of who gets to create, to consume, to judge, and the social contingencies undergirding all these privileges.

Arthur C. Danto writes about the 'transfiguration of the commonplace, banalities made art' (1983: v), but the aesthetic relationships between performance, banality and transformation are not one way. They are, more accurately, circular. The commonplace is created as the operating condition of performance as surely as it may appear trans-figured on the stage. As individuals and communities labor in service of the true, the good or the beautiful in performance, performers remake themselves, often literally, corporeally. But more than this, they reshape possibilities for intimacy. They alter time and space, regulate or recon-ceive gender norms, fashion ways of entering or evading personal and cultural history. They tactically deploy the transcendent and the inef-fable to act out, or dance with, the contingencies of the here and now.

Dance technique puts aesthetics in motion. It is the primary tool by which ideals are incarnated or resisted. The more explicitly physical the aesthetic domain, the more powerful the imprimatur of technique and the more forceful its vocabulary, rhetoric and rituals. If performance is the ideal lens through which to view the social lives of aesthetics, then dance brings these lives into focus with special clarity. For dancers, 'time-less' aesthetics are distilled into the routine, daily discipline demanded by specific techniques.

Technique is both the animating aesthetic principle and the core ambivalence housed in every dance studio and manipulated by every teacher, every choreographer, every performer. It is both taskmaster and mastered, both warden and liberator. It demands to be replicated even as it asks to be exceeded. In ballet or in butoh, technique, like language, reaches out to meet us as we are birthed into dance. However pliable and enabling, technique, like the aesthetics it enables, is a pre-existing conversation between bodies, history and desire. Yet it is also true that however 'monumental,' however rigorously drilled into generations of little bun-headed bodies doing *tondues* at barres, both technique and aesthetics are, finally, local. Both are uniquely affected and inflected by the exigencies of those who keep them in play, set them to music, or deploy them to fashion both the sublime and the banal.

This book is about how dancers use aesthetics, and particularly tech-nique, to build diverse and compelling communities within the larger global city. Some of these performers have forged careers in these creative communities; some have found alibis, or refuges or intimacy in them. For some, dance technique is a welcome respite from care and constraint; for others, it is a matter of life and death. For all of them, rela-tionships to aesthetics, learned and demonstrated through relationships to technique, are crafted through daily labors, with the physical often being the least of these. There are labors of employment and recreation certainly, but more foundationally, the labors demanded by technique are affective and relational. They are social, political, spiritual –labors of love and labors of Sisyphus.

In critical terms, 'technique,' like 'aesthetics,' is a useful synecdoche for the complex webs of relations that link performers to particular subjectivities, histories, practices, and to each other. The threads of these webs are various and teasing them out offers a useful overview of the complex connections between dance technique and the construction of communities. First, and most obviously, technique inserts its object-bodies into language, offering a common idiom through which these bodies are examined, described and remade. Technical idioms make communal aesthetic judgments possible precisely because they organize socialities around these lexical, descriptive endeavors. Thus, technique is, simultaneously, a lexicon, a grammar of/or affiliation – even a rhetoric – in motion. It facilitates interpersonal and social relations as it shapes bodies.

Relationships between corporeality and language are sometimes represented as especially, even uniquely fraught, with dance serving as a special limit case. But this line of reasoning ignores the practical reality that all performance, including dance, is enmeshed in language, in reading, writing, rhetoric, and in voice. Steps and positions have names; there are syllabi, including standardized written ones, for training programs. Movements tell stories, sometimes abstractly, sometimes mimetically and, further, movement is taught through stories, both sacred and banal. The generative imprecision of metaphor is used to communicate how a movement looks or feels; occasionally these metaphors are actually linked to the graphic, as when young ballerinas are told to trace circles on the floor with pointed toes as they learn *ronds de jambe a terre*. Dance has inspired its own graphesis, Labanotion being the best known of these; some such systems emphasize decontextualized movement analyses, not unlike positivism in appeals to rigor and reproducibility.[4] But there are also the reviews, press kits, grant applications, the textual elements central to an artist's or company's professional survival, all requiring that moving bodies be inserted into reading and writing, however uneasy or complex the process.

The performing body's relationships with reading and writing have inspired two textual operations of my own here: one theoretical and one representational. In Chapter 1, I borrow and adapt Robert Scholes's formulation 'protocols of reading' to introduce and describe the ways technique shapes its object-body, making it available for conversation, and actual reading, in a community organized and sustained by a shared idiom. Scholes's protocols point toward understanding, interpretation and criticism: stances and operations to adopt when reading books or bodies. My use of the 'protocols of reading' – and writing – though elastic, is related. Here protocols of reading and writing are first generated by dance technique itself; they are activated as they are incarnated in the classroom, rehearsal hall, and in performance. They are codes governing and standardizing dance practice: where the 'center' of the body is, core movements and positions, norms that cannot be violated. They determine how members of communities organized by technique analyze and understand their re-written bodies. But further, these protocols can be reactivated by audiences to decipher performing bodies. They make these bodies legible and intelligible, and offer bases for interpretation and critique. A given technique's protocols for reading and writing bodies are where the production and the reception of performance meet: in and through cognitive, organizational and relational operations and rubrics binding dancers to audiences and in common vocabularies, grammars, and pedagogical and interpretive practices.

Reading and writing are not the only discursive dimensions of performance technique. It is also suffused with voices: of teachers counting beats or offering praise or correction, of students complaining and complying, of gossip and reminiscence, tall tales and wishes. Technique's protocols for writing and reading the body organize voices as well, though these, like all social aspects of dance, are at least as likely to exceed or circumvent as they are to be contained by such rubrics. Barthes wrote, 'The "grain" is the body in the singing voice, in the writing hand, in the performing limb' (Barthes, 1985: 276). I've attempted here to represent, however inevitably imperfectly, the grains of voices in dance technique and in the communities it generates. This book is filled with voices. Many are dancers who, in their own idiosyncratic ways, offer something akin to the 'murmurings of the everyday' (Mayol, 1998: 7), the routine labors and conversations demanded by the medium. Others are those of dance and performance scholars and critics. Still others are those of artists and theorists not typically brought to bear on discussions of dance or performance. My goal here is, in part, to adopt the evocative, affective, citational processes Della Pollock sees as central to performative writing and criticism (1998). This is the closest I can come to making writing dance. At the least, I hope to approximate the myriad daily partnerships of dance and discourse as these cut across theory and practice in movement's social lives.

The voices of and in technique are not only those of the here and now; they are also those of there and then. Protocols for reading and writing bodies look forward and backward at the same time, to performers who might be and performers who were. Technique, in this view, is an archive. To posit this is both to acknowledge the historical embeddedness of dance and, further, to realize, after Derrida, that 'the question of the archive is not, we repeat, a question of the past . . . It is a question of the future, the question of the future itself, the question of a response, of a promise and of a responsibility for tomorrow' (1995: 36).[5] Certainly for the Sem family, who attempted to use Cambodian classical dance to answer back to the Khmer Rouge autohomeogenocide and the privations of refugee life (Chapter 3), technique as the archival promise of, and responsibility for, the future is deeply and painfully literal. But, while the psycho-emotional stakes are lower, this is also true, and true in reverse, as a promise of inserting oneself meaningfully into the past, in less fraught scenarios: for the adolescent ballerina who calls barre work 'meditating' because 'it's history, it's not personal'; for another who, on completing multiple pirouettes, beams and proclaims, 'Pavlova was here!' and for her partner who replies, sardonically, 'Invite her to performance day, too.'

In 'Thirteen Ways of Looking at *Choreographing Writing,*' Peggy Phelan writes:

> [I]n its journey from disappearance to representation, the body does not 'belong' to the subject who wears it, who dances in and through it – Rather, the represented body comes into Being as it is apprehended within the frame of the representation. In that apprehension, the prop of the body . . . is seized/seen and taken *to be* 'the body.' These props in dance are called 'movement signatures . . . '
> (1995: 20–5)

Or, I argue, they are called technique. I would put Phelan's observations another way. I would suggest that the strategic, 'theoretical' representational frame of technique renders its real-time incarnations both immediately readable and archival, that is, repeatable, memorable and, ultimately, 'theoretical' again. Sometimes representational, technical frame and its incarnation perfectly cohere and sometimes they do not. 'Pour yourself into the movement,' one of my ballet teachers used to say; 'Spilling, spilling,' was her correction. Or, as Xavier Roncalli of Le Studio observed of colleague Roberto Almaguer's flawless technique, 'He *wears* those turns.'

Phelan uses the term 'wear' as well, and the notion of 'wearing' clarifies some of the deeply subjective elements of technique as communal archive, for as surely as it may be soothingly 'not personal,' technique, in its historical, incarnational peculiarities, is also highly individual. One of these personal elements is a given performer's genealogy: Cecchetti technique or Russian? Butoh with Hijikata, Ohno or Tanaka? But more than this, performers wear technique, whatever their genealogical particulars, like clothes – not like costumes or props, though this may also be true – but like the clothes of loved ones who have passed on.

In his moving essay 'Worn Worlds: Clothing and Mourning,' Peter Stallybrass reflects on wearing the jacket of his late collaborator Allon White. 'Bodies come and go,' he observes. 'The clothes that receive those bodies survive' (Stallybrass, 1999: 29). So it is with technique, which likewise survives the bodies which enact it, even if the enactments themselves are always and inevitably disappearing. Sometimes technique is an unwanted and burdensome legacy to be resisted; at others it is a generative inheritance. Either way, '[a]s it exchanges hands, it binds people in networks of obligation' and permission, inextricably linked to memory, both personal and institutional' (Stallybrass, 1999: 30).

The archives of technique, like the closets filled with the clothes of a dead relative, are haunted places that beckon with the promise of rediscovery or beg for exorcism. Avery Gordon writes of haunting as '. . . a paradigmatic way in which life is more complicated than those of us who study it have usually granted. Haunting is a constituent element of modern social life. It is neither premodern superstition nor individual psychosis; it is a generalizable social phenomenon of great import' (1997: 7). To stand in a dance studio is to occupy a haunted place and to write about what happens there is to confront a community's ghosts at every turn: company members, now gone, who became 'sad cases'; affairs and dalliances not to be talked about; fellow dancers who have died of AIDS; crippling depressions; physical breakdowns. There is enchantment too, of course, and fully as much, but whatever its emotional valences, the haunted studios index their ghosts by things like good feet or bad turns or the breathtaking backbend – corporeal, technical madelienes. The specters are prospective too: pointe shoes waiting at the horizon of little ballerinas' practice, choreographic opportunities to make the shades dance to your own tune. And there may be ghosts more traditionally defined afoot as well. For the young ballerina quoted above, 'Pavlova was here.' For my first ballet teacher, in her dusty tile-floored studio in Detroit, it was Pavlova's *absence* that did the haunting. 'If only,' she would sigh, 'if only Pavlova would walk into my studio.' For Ben Sem, the thirteenth century Apsara and the spirits of murdered colleagues seemed both constant goads and relentless reproaches. Oguri engaged Poe's 'The Black Cat' for a turn in a conversation with butoh that began on his mentor Min Tanaka's farm. So thick with both dreams and burdens are the haunted communal archives of technique that I think of them as hauntopias.

These hauntopias may be as diffuse as an ensemble's collective memory, or as condensed as a piece in the repertoire. Technique constructs and occupies both affective and literal landscapes. It activates chronotopes of home and of the road, linking location to time and desire; it performatively instantiates place in space as it maps bodies, rooms, stages and relationships. In Chapter 2, I discuss technique's performative protocols for organizing space and time at length, but the larger geography of this book bears mention here.

Dancing places, placing dances

The techniques I explore in the following chapters include ballet and butoh, Khmer classical dance, Pilates training, and modern-postmodern fusion; the sites I examine include professional and semi-professional companies, established schools and studios, individuals and families. But all reside in the Los Angeles area, including adjacent cities like Long Beach and Pasadena.

Los Angeles is the United States's second largest city and its largest metropolitan area. Janet L. Abu-Lughod labels it one of America's three 'global cities,' marked by 'an expansion of the market via the internationalization of commerce, a revolution in the technologies of transport and communications, the extensive transnational movement of capital and labor, a paradoxical decentralization of production accompanied by a centralization of control over economic activities, . . . the increased importance of business services . . . [and] a presumed bifurcation of the class structure . . . ' (1999: 2). But it is also 'global' in a much simpler demographic respect. According to the Brookings Institution report, 'Los Angeles in Focus: A Profile from [US] Census 2000,' Los Angeles has the second highest proportion of foreign-born residents of the 100 largest US cities; only Miami has a higher percentage. By 2000, immigrants made up 40 percent of the city's population. Two-thirds of this immigrant population comes from Central and Latin America; one-fourth comes from Asian countries. Immigrants to the Los Angeles metropolitan area come from (in descending numerical order) Mexico, El Salvador, Guatemala, Philippines, Korea, Iran, Armenia, China, and Vietnam. In its demographics, and in its physical, social and economic geography, Los Angeles is simultaneously a set of interconnected ethnic and immigrant enclaves and one big metropolitan contact zone where 'the concrete predicaments denoted by the terms "border" and "diaspora" bleed into each other' (Clifford, 1997: 246–7). These predicaments – intercultural collaboration, nostalgia, community building, longing, loss and cross-cultural (dis)identification – are all taken up in the pages that follow.

Los Angeles's status as a global city locates it at the intersection of two trajectories in urban studies. It is both an exceptional city where the local is uniquely subject to 'complex practices of identity and community formation within broader networks of globalization' in the United States and beyond, and a paradigmatic one, 'exemplary of the tendential urban processes at work in all "world cities"' (Villa and Sánchez, 2004: 499). *Dancing Communities* reflects both of these trajectories. Here, Los Angeles is a unique laboratory for examining the complex communal arrangements enabled and sustained by performance, and a paradigm case for a larger argument about the social and aesthetic force of performance as urban political infrastructure.

Henri Lefebvre writes, '. . . detestable cities are also fascinating, for example, Los Angeles' (1996: 208). Dancers say the same thing. As Valerie A. Briginshaw writes, 'The particular ways in which cities and subjects "mutually define" each other are evident when intersections of dancers with urban landscapes are examined. Bodies and cities can be seen to ' "inscribe" each other' (1997: 36). Two crucial aspects of Los Angeles are central to these mutual processes of definition and inscription: one based on what Los Angeles is and the other on what it is not.

Los Angeles is not New York. For dancers, more than other performers, this distinction is especially important. There is no explicit, public dance infrastructure, no major performance series, no large, nationally known resident companies. Dance Kaleidoscope, a highly regarded summer festival with extensive coverage and reach, served as a showcase for artists and companies across Southern California until it folded with the retirement of its indefatigable leader and animating spirit, Don Hewitt, in 2001; there was no bench-strength – fiscal, logistical or creative – to replace him. Los Angeles is, as Norman Klein observes, 'the most photographed and least remembered city in the world,' and its own 'erasures of memory' include its dance history (1997: 247). The dance scene in Los Angeles seems perpetually borne yesterday. It was not. The pivotal Denishawn school and dance company were founded here in 1915. The 'First Negro Classic Ballet' was established in Los Angeles in 1946.[6] Yet as surely as there is no cultural infrastructure in the form of large key companies or events, there is no infrastructure of memory on which dance in Los Angeles builds. New York beckons like a siren song to committed dancers; as one dancer-choreographer explained to me, her decision to relocate there meant 'I was serious as an artist because there I'm part of a serious world.'

Los Angeles is an industry town, *the* Industry town where 'the Industry' emphasizes the manufacture, indeed mass production, of entertainment product: by definition, *not* a serious world. The commercial imperatives of 'the Industry' cut two ways for performers. There are a lot of relatively well-paid, if temporary, jobs that are not dependant on the hand-to-mouth grant pittances sustaining 'serious artists.' Yet these same commercial imperatives circulate like aesthetic plasma through the dance community; perhaps no other city except Las Vegas requires performers and choreographers to explicitly reckon with commercial aesthetics, and affirming or rejecting them, in their work.

Compared to New York, and perhaps only by that standard, rents in Los Angeles are relatively cheap; this is largely a function of its spatial 'patterns of fission' (Abu-Lughod, 1999: 359). Somewhere, in a neighborhood theatre in Venice, a women's shelter in South Central, or an eastside urban university studio, class and rehearsal space can be had without extreme sacrifice. That said, two other commodities make the access to these spaces a complex proposition. Performers need cars and commuting time which, in turn, require some means of support sufficient to finance the car, classes and accoutrements that is also flexible enough to accommodate class, rehearsal and performance schedules.

Dance in Los Angeles is both a large world and a small one. It is large, not only in the sense of being geographically dispersed but in the types of dance available: folklorico of every type; Japanese, from butoh to bugaku; Philippine, Khmer, Korean; classical dance from ballet to Bharata Natyam. The city's artists, including those in this book, globalize the local and localize the global as matters of individual identity and aesthetic practice. They come to Los Angeles from Mexico, Korea, Vietnam, Cambodia, Japan, Cuba, Columbia and East Los Angeles, Pasadena, and Long Beach, to name only a few. These dancers and choreographers engage in complex conversations with classical and traditional techniques, as well as with commercial and contemporary elements, generating compelling fusion and hybrid idioms: corporeal and performative 'thirdspaces.'[7] Yet it is also small in the sense that, for a highly dispersed community, one, moreover, lacking a stabilizing core performance event like the now-defunct Dance Kaleidoscope, there is also a considerable degree of interconnection, both in professional allegiances and in personal relationships. This small world came up in my research. One of Oguri's and Roxanne Steinberg's (Chapter 1) core collaborators also danced with Hae Kyung Lee (Chapter 4). Roxanne's sister was well acquainted with one of the men who trained at Le Studio (Chapter 2). These professional and interpersonal small worlds cross the dance map of the city. Further, they have their analogues in the tactics dancers use to remake Los Angeles through the imperatives of their given techniques. Time after time, I was told that 'everyone' knew the warehouse store with the best deals on dance clothes and shoes; 'everyone' knew *the* acupuncturist or chiropractor who understood dance injuries; 'everyone' knew who offered the best floor barre, or Sunday class, even though 'everyone' may have lived considerable distances from these prime providers. This kind of dance 'folk geography' of Los Angeles remaps and domesticates this sprawling global city, dissolving municipal boundaries in favor of others based on technique in community. And, of course, despite what 'everyone' knows and where 'everyone' shops or trains, it is also possible for dancers to be utterly and completely isolated by circumstance or by choice.

Dancers' social, relational maps may rewrite Los Angeles's literal landscape as a series of connections between cheaper shoes, good teachers and understanding healers, but the social possibilities (or the isolation) organized by technique often exceed material space. Dance technique invokes, organizes and reproduces affective and metaphysical terrains as well. In her novel *A Feather on the Breath of God*, Sigrid Nunez writes of ballet class as her narrator's refuge from her difficult childhood in a Brooklyn housing project: 'But above all else, ballet meant escape. Instead of going home after school, I could go to class. . . . I had discovered the miraculous possibility that art holds out to us: to be a part of the world and to be removed from the world at the same time' (1995: 100–1). Dance as an unqualified, private metaphysical reverie is a rare privilege; as the Sem family demonstrates in Chapter 3, 'aesthetic' remove from the material world is not simple and the costs not easily calculable. As Nunez implies, the escapes technique offers are never unqualified; they are always contingent upon the dialogic interanimations of individual and company exigencies, material constraints and opportunities, affective longings and compromises.

Joyful attachments

In her book *The Enchantment of Modern Life*, Jane Bennett writes: 'In [a] cultural narrative of disenchantment, the prospects for loving life – or saying "yes" to the world – are not good. What's to love about an alienated existence on a dead planet? If, under the sway of this tale, one does encounter events or entities that provoke joyful attachment, the mood is likely to pass without comment and thus without more substantial embodiment' (2001: 4). Dance technique offers precisely this embodiment of joyful attachment, a capacity to be enchanted, 'struck and shaken by the extraordinary that lives amid the familiar and the everyday,' even within the social and commercial flows of the global city (Bennett, 2001: 4). These opportunities may reside in discovering physical abilities individuals never knew they had. They may appear in dance's capacity to channel what Randy Martin has called 'the multiple forces that give rise to mobilization' (1998: 14) of, and in, communities across multiple dimensions of difference. The enchantments of technique are pregnant with the possibilities of provisional utopias, though these are not guaranteed. In practical terms, technique provides social bedrock for imagining new ways of being together and being oneself. In Hae Kyung Lee's company, discussed in Chapter 4, new ways of being include an ethic that bridges divine and daily space with nonprescriptive rituals of care and 'response-ability'. This ethic sustains members through daily and public triumphs, as well as personal tragedies. These are not matters of mindless optimism or naiveté. As Bennett so bracingly demonstrates, enchantment is not vacuousness or vapidity (2001: 14); Hae Kyung Lee and her dancers are not 'typical' La-La-land new age subjects, if, indeed, there are such things. On the contrary, dance's communal enchantments are intimately linked to technique's lexical function of making the body legible in a shared idiom. They are ways of balancing the seemingly magical wonders and potentials of corporeality with its limits, ways of rewriting oppressive relationships to (in)visibility: strategies of solidarity in difference.

These accounts of dance's social lives span 14 years, roughly from 1990 to 2004. Some of the artists presented here are nationally and internationally known professionals. Others are, or were, semi-professional, students, or outside conventional categories of pedagogy and ambition. In these latter cases, I have used pseudonyms, with the specifics of use dependent on individual circumstances as detailed in the relevant chapters. Pseudonyms are designated by quotation marks, except in Chapter 3 where, for reasons detailed therein, they are used throughout. All of the artists whose real names appear here, and whose words and work constitute the heart and soul of this project, have been given copies of the chapters in which they appear and encouraged to provide feedback.

Although a wide range of techniques, ages, ethnicities and social conditions are included here, I have not attempted to construct a 'representative sample' of Los Angeles dance or dancers. I have written about individuals and companies I was drawn to because I was compelled by the communities they created in their art and in their lives, by their aesthetics or their characters, or by their personal and creative dilemmas and tactics for managing them; usually it was all of these. In some cases, I had the good fortune to train with these artists in their specific, technical idioms. In other cases, I drew upon years of observation, informal conversations, and formal interviews. In most cases, I relied on all of these. Formal interviews are designated by their dates. All other quotes are taken from informal conversations and field notes.

The urban communities forged from dance technique are intimate ones, suffused with pleasure, propriety and the play of the vicarious. As one of my long-time informants put it, 'Take a bunch of sweaty bodies. Put them together day after day and, well, you know.' These intimacies include individuals' fantasies and interpersonal conversations; they are the social and affective micropractices from which complex communities are built. Chapter 1 takes a detailed look at the small intimacies that make dance communities function off- and onstage. The chapter begins with a discussion of how technique, as a set of protocols for reading and writing bodies, makes interpersonal conversations about those bodies possible. Sometimes such conversations are intimate in oblique ways, using tactics laden with the possibilities of their own disavowal. Sometimes these conversations are fantastical, exposing the vicarious pleasures undergirding the social aspects of spectatorship, particularly in cases of virtuoso performers like butoh dancer Oguri.

Chapter 2 examines how ballet creates communal spaces and times through the daily routines required to produce it. In a Pasadena ballet studio and its affiliated semi-professional company, these collective chronotopes organize more than schedules and stages. They construct social landscapes offering safety or exposure, competition and cooperation. These vernacular landscapes of ballet are as gendered as its repertoire. For young women in this school and company, ballet becomes a place of refuge and a site of proto-feminist solidarity. For men and boys, particularly heterosexual ones, any such refuge is inevitably qualified by ballet's exposure of the plate tectonics supporting heteronormative masculinity.

Not every urban sociality forged through dance is generative or successful. Chapter 3 examines the fraught intersections of technique; memory; and Bahktinian answerability, the imperative to answer with one's life for one's art, for a Cambodian refugee family in Long Beach, California. For the Sems, survivors of the Khmer Rouge's killing fields, these intersections were particularly perilous. Ben and May Sem tried to use classical dance to legitimate their own survival, and to answer back to atrocity with profound discipline and transcendent beauty. For a variety of reasons, this strategy proved to be inextricably bound to crises of truth, poverty, and social and psychic dislocation.

In Chapter 4, I explore technique's potential to anchor a communal ethic in everyday practice across multiple dimensions of difference. Hae Kyung Lee and Dancers, a Los Angeles-based multi-ethnic, contemporary company, has bound technique to a non-doctrinal ethics of service to each other and to a larger vision of the power of dance to remake their lives and the world. This vision circulates through their performances and their daily routines. It sustains them through physically challenging choreography, overt and passive aggressive racism, and communal grief.

Finally, in my conclusion, I describe dance, and the social arrangements it organizes, using Berlant and Warner's formulation 'queer intimacies.' I argue that the queer intimacies of dance offer its constituents more inclusive, complicated and productive engagements with urban environments. This is especially important to consider at a moment of declining arts funding at all levels. What would respect for queer intimacies offer to urban policy and analysis? I argue that such respect would generate a robust, vernacular sense of the city as a 'place of meaningful proximate links' *and* as 'a site of local–global connectivity' (Amin and Thrift, 2002: 27).[8] In this view, social and political macro-landscapes are inextricably linked to daily, ethical micro ones through the diverse bodies who dance together through them all. This perspective would generate a renewed commitment to, and respect for, both amateur arts practice and performances in public venues. It would take the notion of dancing the city seriously and literally.

In her essay 'Oedipus Rex at *Eve's Bayou* or The Little Black Girl Who Left Sigmund Freud in the Swamp,' D. Soyini Madison, writing in the persona of 'Noir,' argues: 'Creative acts and the talk about them are undeniable change-makers, carrying more pleasure than pain, at least for me. That is why we can't live without them. That is why all the honest words about them are like rain on fertile ground. You see something created, you hear it, and it stays with you, for better or worse. It is effective, for you; and, it affects you. It becomes a political and emotional encounter. It refuses silent appreciation, because it is not simply appreciated; it is more complicated than that' (2000: 315). I offer honest words here, but no simple 'whole truth.' This is not just a matter of the inevitable partiality in all situated knowledges. Sometimes membership in a community means putting one's pen down and keeping one's mouth shut. I have been asked to keep some secrets about aspects of the socialities organized by the techniques discussed here and I have kept them. In some cases, I was not asked explicitly, but I have kept those too.

The works and the talk I have shared with the artists discussed here have affected me profoundly and, as Madison states, simple silent appreciation does not suffice. These chapters, across their specifics, argue against the notion that performance, and dance in particular, happens only to disappear. Performances persist in minds and hearts, in places and in talk, as surely as they do on videotapes. They persist because, as their colors and the edges of their images blur and run together in our minds, they bleed into other images – those of partnership, of family, of creative possibilities and prohibitions, of the global city itself – all the effective and affective elements that make urban life, and art, go. These chapters argue for the power and persistence of performance, not only in such images, but also in the banal, everyday routines that surround their production and reception in complex communities. Technique is the infrastructure that binds aesthetics and these persistent excesses of performance together, linking the powerful, resonant moment onstage to the daily labors in class and rehearsal. It stands inanimate, waiting for urban artists and audiences, with their fantasies and peculiarities, their blessings and their baggage, to make the first, and then the next, and the next move.

1
Intimacies in Motion

*To be As One with someone, whether to music or to movement, is
what I anyway have always wanted – and all the morals and tenets
of Western civilization that try to convince us there are higher things,
they've always seemed to me just a handful of dust. But then, I'm
from L.A., and I have always been what educated people from the
East Coast or England take one look at and think is what's wrong
with this place.*

(Eve Babitz, 'Bodies and Souls', 1994)

*Affective life slops over onto work and political life; people have key
self constitutive relations with strangers and acquaintances; and they
have eroticism, if not sex, outside of the couple form. These border
intimacies give people tremendous pleasure.*

(Lauren Berlant and Michael Warner, 'Sex in Public', 1998)

In their analysis of the global city, Amin and Thrift observe: 'There is a
politics of sociality that needs to be revisited . . . [F]riendship has become
an even more important part of the urban social fabric, many of whose
pleasures lie in simply relating to others – in part as a result of the
increasing emphasis on relationship as a value in itself' (2002: 158–9).
In dancing communities, politics of sociality, including friendships, are
set in motion by myriad daily practices which serve as rhetorical and
corporeal tools for interpersonal and intercultural communication and
cooperation. These operations, in turn, organize complex, heterogeneous, productive social formations onstage and off, including 'what we
consume [and] what we see' (Amin and Thrift, 2002: 137).

This chapter examines the interpersonal micro-practices that enable
and sustain friendships in dancing communities. These communities

are not made by sylphs, spirits, automatons or pure physical potential, despite their prevalence in the repertoire. To the contrary: real, desiring, emotional subjects come together across multiple dimensions of difference to produce and consume dance because doing so offers real, emotional pleasures. Intimacy is foremost among these. It is both a condition for, and a by-product of, the sustained physical and affective engagements central to the complex social arrangements in dance. Real and vicarious intimacies create and expand possibilities for community within the global city through the accumulation and deepening of interpersonal exchanges.

Lauren Berlant and Michael Warner argue that '[c]ommunity is imagined through scenes of intimacy' (1998: 554). In the multi-ethnic dancing communities I discuss here, these scenes are 'local, experiential, proximate, . . . saturating' (Berlant and Warner, 1998: 554 n. 15). They operate both within and outside of conventional, culturally specific romance plots of coupling, kinship and domesticity.[1] They link private relations and fantasies to the social work of performance.

How are these intimacies organized? Part I of the chapter examines the social and aesthetic bedrock that makes intimacies in complex urban communities possible: dance technique. Robert Scholes's formulation 'protocols of reading' serves as a tool to explicate the lexical, grammatical, social functions of technique. Technical protocols make intimacy possible by offering shared vernaculars and interpretive strategies; these, in turn, support the interpersonal and communal exchanges that make dancing communities go.

Next, I turn to two sites that challenge conventional presumptions about intimacy even as they expose, in unique ways, the micropractices making it possible: Pilates training in Part II and spectatorship, specifically of virtuosic performances, in Part III. In one respect, these sites exemplify the most banal operations in the urban landscape: consuming, with its attendant routine conversations, and looking. Each is, on the surface, easily reducible to simple politics of propinquity, exchanges of/in difference that, while providing texture to city life, are non-transformative. Yet the relationships that develop in the specific sites discussed here, between Pilates instructors and their clients, between performer and audience, offer interlocutors much more.

Pierre Mayol observes of related urban transactions, 'One could not just remain in a simple relationship of consumption with them because the relationship became – it had to become – the support for another discourse that here, in a generic fashion, I call confidence' (1998c: 83). Mayol's 'another discourse . . . confidence' is one kind of intimacy and, while the two sites discussed here are interesting laboratories for examining its formation, they differ in both the kind of intimacies offered and in the logistics of the offerings. Pilates training at Le Studio Fitness, a Pasadena training facility, enables close scrutiny of the talk and touch that link bodies in motion to one another over time. Butoh dancer Oguri illuminates the looking and longing, the projections and play of the vicarious, that bind audiences to exceptional performers dance after dance, year after year. Both rely on specific technical protocols to birth other discourses, real and fantastic, across the boundaries of seemingly conventional arrangements. Both rely on operations that animate technique as a matter of daily practice, including the generative instability of metaphor, the enabling discipline of propriety, and the play of the vicarious.

Taken together, these same practices and operations link the dynamics of dance's communal production to those of its communal consumption. They link alliances forged in the routine work of training to those forged by spectatorship, and bind the banality of daily interpersonal exchanges offstage to the transporting effects of witnessing virtuosity on it.[2] They remind us that the most ordinary intimacies in performance, and the most extraordinary ones, happen in interpretive communities constructed by shared protocols that make such intimacies communicable to the wide range of constituents on stage, in the studio, and in the audience.

PART I   PROTOCOLS FOR INTIMACY

Dance technique is relational infrastructure. It offers templates for sociality in the classroom and in the performance space. Technique translates individual bodies into a common 'mother tongue' to be shared and redeployed by its participants: a discursive matrix, a vocabulary and a grammar, to hold sociality together across difference and perpetuate it over time. At its most basic level, technique births new templates for sociality by rendering bodies readable, and by organizing the relationships in which these readings can occur.

Robert Scholes's formulation 'protocols of reading' (1989) is especially useful for describing these organizational and relational functions of technique. Though his focus is on conventional literary texts, 'protocols of reading' is also applicable to embodied practices. Not only are such practices, and dance in particular, utterly enmeshed in textuality – from syllabi and lists of exercises to manifestos, trademarks and press kits – but Scholes's view of reading also emphasizes the highly situated nature of textual encounters: their materiality and contextuality. Thus, 'protocols of reading' avoid simplistic text/performance binaries and, in Scholes's applications, function more like Diana Taylor's scenarios (2003: 28).[3]

For Scholes, protocols of reading contain and organize critical, interpretive encounters with texts in both homogenous and diverse communities. They illuminate the flows of power and pleasure, and their points of divergence and convergence, as these are encountered by students and teachers through the social processes of reading (Scholes, 1989: 151). In a similar vein, dance technique generates metaphors and models used by dancing communities to organize the powers and pleasures of rendering the dancing body intelligible and communicable.

To reconceive specific dance techniques as protocols of reading is to take them out of the realm of objects dancers have and then deploy. Instead, techniques are recast as sets of communicative and interpretive conventions shared by performers and audiences who, both separately and in concert, produce and consume the dancing body. Dance technique, in all its forms, is constitutive in its ambitions, but its protocols offer communal preconditions for, not limits to, its social uses. Thus, these protocols generate collective criteria for producing, interpreting and evaluating dancing bodies, for 'reading,' 'writing' and speaking them, even as they highlight the contingencies involved, including economies of discipline and of pleasure.

Two additional elements of Scholes's protocols are especially useful for exploring how technique organizes dancing bodies into readability and relationships. First, 'protocols of reading,' like technique itself, circumvent a theory/practice dualism, as well as its subtler manifestations in which pure technique/theory is an end in itself and performance/practice simply its illustrative, exemplary servant or means. Scholes observes, 'There is no place outside of practice where theory might stand in order to dictate protocols of reading... But there is no place outside of theory for practice to stand either' (1989: 88). Likewise, Susan Leigh Foster argues for two inter-animating bodies created by technique, 'one, perceived and tangible; the other, aesthetically ideal' (1992: 482). Thus, technique, as a set of protocols for mapping and reading the body, is simultaneously constituted by an overarching, ideal vision of the subject-ed body, and through the micropractices which actually inscribe this vision onto specific bodies with varying degrees of success or failure.

Second, protocols of reading, like dance technique, are not totalizing, whatever the nature of their claims. Their very materiality precludes this. As Scholes observes, 'Texts and people [and bodies] do not abide in some timeless moment but in time. They are thoroughly impregnated with time; they are constructed and deconstructed in time . . . And nothing made of time and functioning in time can be complete or perfect' (1989: 151). As Foster observes of the relationship between the futile, total-izing claims of technique and the materiality of bodies which instan-tiate these over time: 'The prevailing experience . . . is one of loss, of failing to regulate a mirage-like substance. Dancers constantly appre-hend the discrepancy between what they want to do and what they can do. . . . The struggle continues to develop and maintain the body in response to new choreographic projects and the devastating evidence of aging' (1992: 482).

The totalizing ambitions of technique are not only undone by the body's frailties; they are also belied by the body's achievements, as Randy Martin asserts: 'the fundamental contradiction of all technique is that the very training which allows those who participate to be mastered by discipline also allows them to achieve mastery over technique itself. For mastering technique develops a fluency in practice that loosens the fixed hold on the body initially commanded by that very technique' (1998: 20).

Central to this discussion of dance technique as protocols for reading bodies is sociality. For example, both Scholes and Foster emphasize the importance of the dialogic (to varying degrees) classroom as one site where technical protocols for reading books and bodies are incarnated and interrogated (see Scholes, 1992: 78; and Foster, 1992: 482–5).

Extending Scholes's insights to mapping and reading bodies in dancing communities suggests that technique combines elements of strategies and tactics as described by Michel de Certeau. Here, at the most basic level, technique-as-strategy transforms corporeal possibilities into readable, reproducible entities; it 'postulates a place [or a body] that can be delimited as its *own*' (de Certeau, 1984: 36). Tactics, on the other hand, are 'guileful ruse[s]'; what they 'win' they cannot 'keep' (37). They are mobile, alert to 'cracks' in the facades of 'proprietary power.' Thus, dance technique aspires to systematically organize and hold fast to its object – body while, at the same time, it is both improvisatory (depending on the relative condition of that body) and opportunistic (using unexpected abilities/strengths in one area to compensate for and navigate through movements that reveal weaknesses in other areas). As Foster observes, 'the dancer perceives a certain technique for reforming the body and the body seems to conform to the instructions given. Yet suddenly, inexplicably, it diverges from expectations, reveals new dimensions, and mutely declares its unwillingness or inability to execute commands' (1992: 237).

The strategic and tactical dimensions of technique illuminate dancing communities' complex uses of space and time. Strategies, de Certeau posits, organize and secure their places of operation and, in so doing, offer 'a triumph of place over time' (1984: 36). The standard configurations of studio and performance spaces, and the regimentation of dance training, attest to the strategic dimensions shaping the social transmission of technical protocols. This is especially evident in classical dance, where the explicit goal is to (re)create 'timeless' forms. Any triumph over time is highly provisional, though, in part because of the effects of aging on strategic bodies; dancers, like all ' "tactical" practitioners[,] do not have time at their disposal' (Ahearne, 1995: 163).

Tactics intervene in strategies' presumptions of the appropriate use of time in/and space. They divert time, recode space, and insert heterogeneous rhythms into the provinces of strategy for unauthorized purposes. (See Chapter 2 for an extended discussion of tactical, communal space and time in ballet.) Such diversions and reappropriations hide in the light of official syllabi, training regimens, and spectatorial practice, redeploying technique in service of relationships and fantasies that ideal, strategic visions of the body could not, or would not, condone. These relationships and fantasies, flirtations and projections, are enabled by technique's protocols, even if they exceed them. They are affective, relational labors as intrinsic to dancing as the physical labors of stretching and lifting. Like movement with and around other bodies, these labors organized and enabled by technique are practices of everyday urban life. As Celeste Fraser Delgado and José Muñoz observe, they bring 'people together in rhythmic affinity where identification takes the form of histories written on the body through gesture' (1996: 9). Through technique and its tactical redeployments, 'a shifting sense of community is configured and reconfigured – day after day and night after night' (9). Thus, while the protocols of dance technique map and organize bodies, and bind them together in socialities with strategic ambitions, they also serve banal and daily relational duties. They shape intimacies, offer hiding places, and produce modes of reflexivity as they tactically limit or engender forms of solidarity and subjectivity.

Dance technique's strategic and tactical operations are communicatively constituted, whether in the classroom, in performance, or in between. Randy Martin argues that the complex communicative potential of dancing communities is a neglected but useful corrective to key aspects of contemporary progressive social theory: 'However much the importance of subjectivity to account for mobilization was recognized, it remained an interest, an ideational form that could not account for what kept people in motion, for what maintained or mobilized participation. . . . Consequently, the participation generated through the gathering of bodies is undervalued in a theory of politics where putative ends and means are cognitive . . .' (1998: 9). In addition to corporeal solidarity, the vocabulary and tropes unique to various techniques organize dancers and audiences into rhetorical, interpretive communities across multiple dimensions of difference. Yet, while corporeality and common vocabulary figure prominently in the communities generated and maintained by dance technique, it is important to consider that the relationships established by these protocols are also interpersonal, constituted by conversations linking teacher and student, artist and critic, partner and partner. Herein lie the relational pleasures Amin and Thrift (2002) see as central to the politics of sociality in the global city.

To approach dancing communities through everyday talk organized by a given technique means attending to micropractices of pedagogy and performance, and to by-products which may have nothing to do with either. It means taking seriously the accretional build-up of interpersonal and collective intimacies, the 'ephemeral, fragile, and circumstantial' (Giard, 1998: xxxviii) operations characterizing dance as a relational process and a presentational product. Jill Dolan has argued for the utopian potential of theatre, for the generative possibilities of 'moments that work' in performance and rehearsal. She observes: 'My concern here is with how utopia can be imagined or experienced affectively, through feelings, in small, incremental moments that performance can provide' (Dolan, 2001b: 460). These small moments, and their world-making potential, are borne of everyday exchanges enabled by dance's technical protocols for organizing bodies and relationships.

Viewing dance technique as a set of protocols organizing interpersonal conversations does not simply reify or reinstantiate coherent, autonomous, conversing subjects. Rather, this perspective calls for reading and thinking between subjectivities to expose the emotional, relational labors, as well as the material ones, that make multiethnic, multi-dimensional urban communities go. Such a view demands rethinking the very nature of interpersonal communication. As Leonard Hawes has indicated, conversations do not simply reproduce and reconstitute autonomous subjects: 'Conversations foreclose as well as disclose ways of escaping from and relocating to different subject-positions at the same time they redraw ideological boundaries; theorizing conversations in such a fashion renders dominant practices and their transparent codes as audible [and, I would add, embodied and readable] fictions that put into practice novel as well as mundane modes of resistance and surrender' (1998: 273). Further, he argues, '[I]t would be a mistake to theorize conversation as a totality, as some coherent, bounded, unitary phenomenon.... Such discursive micropractices interpellate individuals – as interlocutors – to the speaking voices of performing bodies.... Knowing one's place[s] as an interlocutive subject – staying in it and keeping track of it [or not] – has undeniably real political and personal consequences' (Hawes, 1998: 274–5).

This view of conversation does not assume transparency of consciousness or meaning, or relational symmetry. It acknowledges gaps, and the possibility that seemingly dialogic exchange 'may simply be two people taking turns broadcasting at each other' (Peters, 1999: 264) across gulfs of difference. Further, the interpersonal conversations that make dance go may not even be literal exchanges of words. They may be projective, fantastic, vicarious, what John Durham Peters characterizes as 'a registry of modern longings' and 'those practices that compensate for the fact that we can never be each other' (1999: 2, 268).[4]

The protocols of dance technique make bodies available for conversation and intimacy in strategic places; they may be context or pretext. Susan Leigh Foster has noted: 'when I look at another student in the [dance] class, I see her or his body not as that of a friend or an acquaintance, but as the bodily instantiation of desired or undesired, correct or incorrect values' (1992: 483). Here, her reading of others' bodies, and the sociality re/produced by her look, is strategic. This is not the interpersonal conversational sociality of technique that I mean here, though it may be included. I am referring to conversations that are more tactical and idiosyncratic, characterized by affective investments in which technique may be an alibi and enabler.

In the remainder of this chapter, I turn to two sites of interpersonal intimacy organized by technical protocols: Pilates training, and spectatorship in performances by Los Angeles-based, Japanese butoh virtuoso Oguri. These sites offer opportunities to explore the constitutive potentials of talking, touching and looking in dancing communities. Both produce what Della Pollock calls 'unsteady mutualities' (2001: 204): arrangements especially relevant to the formation of intimacy in the global city. Here people who meet through happenstance, or encounters organized for other purposes, may become, over time, something more. Unsteady mutualities, with their relational asymmetries and ambiguities, may yet birth the solidarities-in-difference so central to these social formations. The following accounts are love stories of sorts, organized by dance technique. John Durham Peters asserts, such accounts are inevitably 'small and partial' (1999: 271), but the larger affective forces at work in these partialities are performance's communal and aesthetic infrastructure. Finding a vocabulary for describing the social work of intimacy in dancing communities is difficult. In her book *I Love To You*,[5] Luce Irigaray proposes a project that, in Della Pollock's words, moves love, 'this most fundamental, shy, constraining and constrained sign of what's left out of the 'political-sexual field' into a felicitous history' (2001: 204).

In its broadest sense, *I Love to You* attempts to bridge the radically interpersonal and the expansively historical and political; Irigaray wrote the work as a consequence of involvement with a debate on 'New Rights in Europe' (1996: 1). Her goal is an approach to relationships in sexual difference that does not reduce the other to object. To do so, she offers a retheorizing of affective life, and of love in particular. Central to Irigaray's project is the relation guaranteed by indirection in 'to,' as in 'I love *to* you,' intimacy, as Pollock observes, 'trembling on a preposition' (2001: 204). Irigaray writes, '*I love to you* thus means: I do not take you for a direct object, nor for an indirect object by revolving around you' (1996: 110). For Irigaray, the interpersonal is a beginning of, not the conventional romantic limit to, larger social elaborations on intimacy and connection.

'To' is important because, to paraphrase visual artist Barbara Kruger, it always hits or caresses some body; it is deployed by, or reverberates between, dancing, observing and writing bodies. This is true in Irigaray's problematic text, where 'to' is sometimes tethered to essential, heteronormative bodies, and sometimes hovering above metaphysical ones. Yet Irigaray's terms, if not many of her presumptions, bring a certain precision to my discussions of intimacy in dancing communities. Here, both Pilates training and virtuoso performance offer those engaging their specific protocols 'exchanges in which the world is born and remains between two bodies, maintaining themselves by respecting their differences and spiritualizing them without removing them from their flesh' (Irigaray, 1996: 125).

PART II   REFORMING INTIMACY IN PILATES TRAINING

> *So language scatters the totality of all that touches us most closely,*
> *even while it arranges it in order.*
>   (Georges Bataille, *Erotism*, 1986)

Valerie Briginshaw notes that bodies and cities 'inscribe each other' (1997: 36). In Los Angeles, this process of inscription is intimately linked, in both the popular imagination and in practice, to 'body work,' perfecting physical appearance and potential in response to real or vicarious engagement with the entertainment industry and its products in ways that cross demographics. Fitness studios and classrooms of all kinds serve as gathering places for those with similar physical ambitions; they offer possibilities for building bodies and/as building relationships, organized by their specific technical protocols. Sometimes, as in the case I examine here, they offer clients and teachers, brought together by relative happenstance, what Mayol calls 'another discourse' (1998c: 83) one that links participants to each another through talk and touch.

For over ten years I have trained privately and semi-privately (with one to three other people) for an hour twice weekly at Le Studio Fitness, the Pilates adjunct of Le Studio in Pasadena, California (see Chapter 2). Like many of the clients, I became interested in the technique when Le Studio owners Charles ('Chip') Fuller and his brother Philip ('Phip') began to pursue Pilates certification;[6] I had taken ballet classes from Chip and Phip for years at Le Studio, observed the school's affiliated company, and saw the technique as an adjunct to ballet classes. One company member, Xavier Roncalli, eventually joined the Pilates studio staff; I worked with Xavier since his arrival. Erica De La O, who was also a company member, and whom I had known since she was a child, also worked at the Pilates studio until joining a professional ballet company. (See her narrative in Chapter 2.) Jeanne King, also from a ballet background, joined the staff in 2001; she and Xavier currently serve as co-directors of day-to-day operations, and supervise the small instructional staff: two or three teachers and occasionally interns seeking clinical hours for certification. Studio clientele, like its instructional staff, is culturally and ethnically diverse. Instructors include first- and second-generation immigrants from Mexico as well as Anglos; clients span a wide range of countries of origin, including recent immigrants from Armenia, China, France, Israel, Korea, Mexico, the Philippines and Taiwan. African American and Latina/-o clients train there as well. The majority of clients are women; men of all ages make up approximately 30 per cent of the clientele and work with both male and female staff.

Most instructors and clients are professionals: artists, dancers, graduate students, small business owners, PR specialists, social workers, teachers, lawyers, dentists. Most of the instructors are middle-aged; clients vary in age from pre-adolescents to octogenarians. Most clients take semi-private classes, which are roughly half the cost of private ones. Both clients and instructors were drawn to Pilates because it is 'dancerly,' linked, particularly, to ballet and modern concert dance.

While a small percentage of clients come and go, most are 'regulars,' coming once or twice a week, often for years. The community is remarkably stable, even to the point of instructors and clients taking on particular affective valences depending on the predictable 'regulars' scheduled on particular days.

The studio is characterized by very close relationships between clients and instructors. Both groups routinely self-disclose in sessions; some socialize outside of class. Over years, this closeness has manifested itself in instructors fixing up clients on dates; treating elderly clients for free, even badgering them to come in if the instructors perceive the need; visiting clients in jail or the hospital; sharing meals; and countless other intimacies. Both groups share personal details about health, relationships, personal plans. The atmosphere, as one client observed, is 'as much hanging out as working out,' though instructors are routinely teased for enforcing discipline during class sessions. Instructors are often theatrical in their teasing of clients, as much a performance to/for colleagues and other clients as relational currency between individuals. Friendships are not only 'vertical,' between clients and instructors, but 'horizontal' as well, linking the instructors and clients to one another.

Mutual support – sharing rides, money, food, tickets to performances, gossip, offering information on health care – seems unusually visible in class sessions, particularly when viewed against the relative anonymity of corporate mega gyms that dominate Los Angeles fitness training. This mutual support is physical as well as conversational, enabled and organized by the protocols of Pilates technique.

Pilates sessions are a lot like Eve Babitz's descriptions of the tango: though they look like the most artificial, most programmatic things men and women could possibly get themselves into trouble doing, actually doing the exercises is demanding on many levels (1994: 128). Pilates training is a system of nonweight-bearing exercises designed to increase strength and flexibility by remapping breathing with intense focus on abdominal muscles. The method was developed by German-born Joseph Pilates; in addition to developing over 600 exercises, linked under a rubric he called 'contrology,' his early experiments with designing exercise equipment for immobilized patients involved attaching springs as a source of resistance to hospital beds. The springs functioned like weights in weight training, but without the concomitant demands on the back. Pilates brought the method with him to New York and opened a studio there in 1926.

Pilates's method, and his equipment, were widely used in New York to rehabilitate dancers. The technique's emphasis on long, lean muscles and on flexibility made it especially sought after in this regard. Still, it remained relatively esoteric until it was popularized as a celebrity fitness workout in the mid to late 1990s. Its core constituency of dancers was a salient feature of its newfound appeal; its ethos was 'cultured' and 'up market' compared to gym-based workouts.

In Pilates training, the student lies on a mat or, more typically, on an apparatus called the Reformer, a sliding board on rollers with straps connected to a variety of springs that can be adjusted to alter resistance (Figure 1.1). Using the abdomen as the locus of movement, the Pilates method encourages practitioners to breathe by keeping the abdomen still, the tail-bone fixed on the floor mat or apparatus, and expanding the ribs laterally. The goal is to have the back supported by the abdomen at all times. Feet or arms are placed in the straps and a series of isometric exercises executed with resistance from the springs. In addition to these isometric, strength-building exercises, the Reformer provides a supportive surface for stretching and increasing flexibility.

Figure 1.1 Interior shot of Le Studio Fitness and Pilates apparatus, 2005. Pasadena, California, USA. Photographer: Charles Fuller

In her book *Samba: Resistance in Motion*, Barbara Browning tells of her experiences as a student of the *candomblé* and *capoeira*; she observes that, to make her body receptive to the *orixás*, and to write about that experience, she tries 'to feel my mind work like a muscle' (1995: 73). In my experience of Pilates training, my efforts move in the reverse; I strive to make my muscles work like my mind. Pilates calls not for abandon, but for a deliberate linearity that is often counter-intuitive. An entire micro-map of the body must be tended to explicitly, beginning with breathing. As noted above, Pilates emphasizes strong abdominals and a concave navel area to support movement generally, and the back in particular. To internalize this element of Pilates' corporeal map, the instructor places a ball on the student's abdomen while she is supine on a mat or the Reformer; a variety of movements are negotiated with the goal of keeping the ball perfectly still. Conventional diaphragmatic breathing is out of the question: the ball pitches and rolls off of the abdomen. Instead, breathing is lateral. The ribs do not move up and down, but the chest cavity expands to the sides while the abdomen stays 'quiet.'

In the most banal terms, the Pilates method can be described as a combination of low-resistance weight training, yoga's emphasis on breath, and ballet's emphasis on stretching with an apparatus that supports the neck and back, minimizing the potential for injury. With the exception of very advanced students, Pilates practitioners are guided by an instructor who has completed a training and certification process, and who serves as a student's individual facilitator, sometimes consistently, sometimes in introductory sessions only.

Several excerpts from a large genre of breathless articles extolling the method in the popular press at the height of its crossover into mainstream fitness, generally written by those who freely deploy a rhetoric of conversion, are illustrative of how the method actually works for someone engaging in the technique. Liz Brody observes: 'When you sign up for a lesson on the Pilates equipment, you generally exercise under the close supervision of a trainer.... On the Reformer alone, you lie, stand, kneel and sit while your body goes through a wide range of motion. Stretching while strengthening, you feel something like candy in a taffy pull' (1998: S1+). Tamala Edwards adds: 'for an hour an instructor leads a client through a volley of positions, both on the floor and on machines with names like the Cadillac and the Barrel. Repetitions are low, but concentration is intense. The stomach and butt squeeze, the legs and arms reach' (1998: 64). Amanda Hesser summarizes the core exercise of most Pilates sessions: 'Alycea took me to the reformer, a long, low, bedlike apparatus with a flat padded carriage that slid back and forth the length of the bed . . . We moved on to more complex movements, my least favorite being the "hundreds." Lying face up on the reformer, I had to raise my feet six inches, suck in my stomach, squeeze my buttocks together and, holding a stirrup set on springs in each hand, keep my arms at my sides and bounce them up and down, 100 times' (1998: D9).

Pilates training, like dance training generally, deploys two Foucauldian technologies: those producing a 'reformed' and disciplined body, adapted to regimes of high culture aesthetics and/or the state, and those of self-fashioning, which, according to Foucault, 'permit individuals to effect, by their own means or with the help of others a certain number of operations on their own bodies and souls, thoughts, conduct, and way of being, so as to transform themselves in order to attain a certain state of happiness, purity, wisdom, perfection, or immortality' (1988: 18). Foucault sees the classical roots of these latter 'technologies of the self' in Stoic techniques of self-examination and self-disclosure, characterized by meticulous concern with reading and reproducing detail, particularly corporeal detail (1988: 28). These classical techniques were framed as relational, involving a mentor/interlocutor. Likewise, the transformative potential of Pilates training is rooted in detailed readings and rewritings of the student body – self-discovery and self-fashioning through submission of corporeal details to the gaze and touch of another. In Irigaray's formulation from *I Love to You*, Pilates turns on a vulnerability to, and a reciprocity of touch, 'a touching which respects the other proffering him/her attentiveness including carnal attentiveness' (1996: 124). I locate the 'to' of 'I love to you' in part in this touch for, in Pilates, it functions as a continual reminder of mutual embodiment, mutual strengths and frailties, a corporeal dialogics attesting to the materiality of the labors of intimacy in dancing communities.

Pilates training is demanding because the requisite breathing is counter-intuitive, the movements are simultaneously minute and taxing, and the sessions are relatively expensive, though less so than sessions with many professional personal trainers. But more: this training demands vulnerability that seems, at first, to be simply biomechanical. Strung up in straps, in positions that evoke images of bondage, the student body is touched in intimate places: pelvis, chest, small of the back. Yet, over time, working with another to read, reread and rewrite the body's possibilities, this biomechanical relationality becomes something else; interpersonally, emotionally, the capacity to be touched becomes the capacity to be moved.

There is an additional dynamic at play in Pilates sessions: that of the classroom. The studio is a training facility for both amateur and professional dancers, athletes, martial arts practitioners, and other fitness students. As Jill Dolan argues in *Geographies of Learning*: 'The classroom is an intimate place. . . . Classrooms are places of longing and loss, in which embodied emotions roil to prompt the pursuit of intellectual fulfillment, a state that can only be attained for a moment. This is the stuff of desire' (2001a: 147). In dance and Pilates studios, as in other classrooms, 'live bodies and minds come into contact to think, to read [other bodies], to analyze together, to perform for each other'; the dance studio is a place 'charged with the mysteries of presence and charisma, with curiosity and longing,' a place of vigorous sociality and 'private passion' (Dolan, 2001a: 147). Here, bodies are reassembled using vocabularies of touch as well as those of language. Here, the student as object-body of technique reconceives her own corporeal geography through the mediation of another who is, simultaneously, an ideal and another body, an interlocutor, a friend.

While all dance technique deploys protocols of reading and writing the body, Pilates training does so in very concentrated ways. Touch and talk conveying these protocols are not diffused across the class, as in a conventional dance studio. Instead, the performing body of the student and the demonstrating body of the instructor are closer, both physically and affectively. As surely as ballet class instantiates the standard repertory, Pilates training births and stabilizes a shared history between its interlocutors: a repertoire of movements and of stories. In this history, and its performances in myriad conversations framed by technique and woven in and through each routine, intimacy is invigorated, revisited, incarnated. Intimacy resides in the accretional, almost geological build-up of shared and specific knowledge of one's physical capacities or limits: what side of the body is stronger; where the scar tissue is; what idiosyncratic abilities specific muscle groups have. These idiosyncrasies are read and later poured into shorthand, into nicknames given by instructors to clients and vice versa, reminiscent of the old neighborhood where familiarity was assumed and protocols of reading also the technology of that familiarity, that intimacy: 'Gumbi-woman,' 'Iron-toe.' There are the stories encysted within these corporeal deficiencies and abilities. Every injury has a story and an attendant invitation to add to the interpersonal and intimate repertoire built up, over years, between instructors and students. This interpersonal repertoire organized by technique is, I suggest, a kind of lovers' discourse in Irigaray's sense, a figural intimacy in Barthes's sense. Consider figures as Barthes does: 'the bod[ies'] gesture[s] caught in action . . . Figures take shape insofar as we can recognize, in passing discourse, something that has been read, heard, felt [or embodied]' (1978: 4), something like the tactical intimacy that insinuates itself into technique's strategic ambitions.

How does such a figural intimacy insert itself into the strategic language of technique, into the rigor of its routines? Irigaray observes: 'If we are to regulate and cultivate energy between human beings, we need language. But not just denotative language, language that names, declares the reality or truth of things and transmits information; we also and especially need language that facilitates and maintains communication. And it is not just the lexicon we are talking about but a syntax appropriate to intersubjectivity' (1996:100). The syntax of intersubjectivity deployed in the figural intimacy of Pilates training, and maybe most dance training, is, as Irigaray suggests, not denotative but metaphoric. Touch and talk, technique as disciplinary project and technique as alibi for intimacy, are affixed to the body by metaphors. Metaphors are the figural, conversational tactics that simultaneously make up and exceed protocols of reading bodies. In so doing, they organize Pilates classes as relational and affective, not simply physical, labor.

Corporeal verities, metaphoric ambiguities

Metaphor is a technology of intimacy, an in-group discourse that creates coherent, if diverse, communities in the here and now, and a set of protocols designed to make images communicable over time. As Ted Cohen observes: 'Three aspects are involved: (1) the speaker issues a kind of concealed invitation; (2) the hearer expends a special effort to accept the invitation; and (3) this transaction constitutes the acknowledgement of a community. All three are involved in any communication, but in ordinary literal discourse their involvement is so pervasive and routine that they go unremarked. The use of metaphor throws them into relief, and there is a point in that' (1978/79: 6). The intimacy produced by metaphor depends on shared relational history and relational labor that must be negotiated by interlocutors to become 'an intimate pair' (1978/79: 7). Further, Cohen notes, 'The sense of close community results not only from the shared awareness that a special invitation has been given and accepted, but also from the awareness that not everyone could make that offer or take it up' (1978/79: 7). These

intimacy- and community-building properties of metaphor are especially acute in dance and Pilates training, and in the affective, relational
labors of instructors and clients that makes training go.

'Throughout any love life,' or, I would add, any Pilates session,
Barthes notes that figures occur without any order, for on each occasion,
they depend on accident (1978: 6). Barthes continues: 'Confronting
each of these incidents . . . , the . . . subject draws on the reservoir (the
thesaurus?) of figures, depending on the needs, the injunctions, or
the pleasures of his image-repertoire' (1978: 6). As Susan Leigh Foster
observes, the needs, the injunctions and the pleasures of perfecting
an action demand, along with example and the intimacies of touch,
the deployment of the array of metaphors that the action inhabits or
evokes (1999: 7). Further, '[a]s the student moves deeper into the study
of dance, the imprecision of the metaphors . . . becomes more apparent'
(Foster, 1999: 7). These metaphors become figural, like the intimacy
they help construct and navigate: 'no logic links' them or 'determines
their contiguity . . . they stir, collide, subside, return with no more order
than the flight of mosquitoes' (Barthes, 1978: 7). For example, Foster
asks: 'In order to align the pelvis in ballet, does the dancer "tuck under,"
"lengthen the tailbone," or "lift from the top of the head, pulled by an
imaginary string"? . . . In executing a Graham contraction, should the
dancer "scoop out the abdomen" or "pull the navel towards the spine"?
Metaphors continually expand and shift as the student learns ever
more nuanced articulations of a given movement system over a course
of study . . .' (1999: 7). Metaphor makes technical protocols for reading
and writing the body communicable, constructing communities of
interlocutors. When metaphors collapse, says Foster, we can no longer
dance (1999: 8).

But, as Cohen noted above, metaphor's utility exceeds the meaning it
conveys. In dance and Pilates training, metaphor makes a double move.
First, it invites its students into the community of initiates. But, in the
intimate labors technique enables and conceals, metaphor gets spun
again. Technique becomes a way of shoring up shared personal histories
as much as shared protocols of reading. Here, over years of working
together, suffusing technique with personal idiosyncrasies of talk and
touch, metaphor offers an invitation within an invitation: a tactic for
staking a personal and intimate claim on apparatus, on movements and
vocabulary, and on the relationships that crystallize around them.

Pilates training is especially metaphorically dense, a set of nested
boxes of metaphors. Not only are individual components of movements
explained metaphorically as in most dance-related pedagogy ('open

your rib cage'; 'spread your vertebrae'; 'keep the navel/shoulders/hips quiet'), but the exercises themselves carry metaphoric monikers ('the mermaid'; 'the tower'; 'the monkey'). Doing a good monkey, Chip says, is the second greatest feeling. The apparatuses have metaphoric resonances – the 'Cadillac,' for example, being the top of the line for sheer theatricality of the exercises performed on it. Xavier once observed, after directing a client to 'do the Teaser on the Table,' that 'someone who didn't know better would think this was a strip club.' Interpersonal dynamics of training itself are explicitly metaphoric as intimacies build up between clients and instructors. One woman called a former trainer at the studio her analyst; another called him her 'past-life husband,' a function, I guess, of their constant arguments. Even frank relational/biomechanical failures turn on the figural pivot of metaphor: 'Your hands are in no-man's land.' 'It's a calf, not a hockey stick.' 'Save the pieces.' 'You're holding me like a football.' These deployments of metaphor tactically domesticate technique while ostensibly communicating its imperatives; they are ways of seizing opportunities for interpersonal play rooted in shared history and shared embodiment.

The imprecision of metaphor both constructs and relativizes intimacy because it is unstable, equivocal, and as Ted Cohen observed above, 'a *concealed* invitation' (1978/79: 6; emphasis added) that may never arrive. At least in one sense, it cannot arrive. Even in the intimate sociality of touch and talk that constitutes dance training, Foster reminds us that metaphor, our technology of community, simultaneously 'teaches most palpably, the slippage between the symbolic and the real' and that 'the "real" in dance, the body, holds in its hands our well-being, our mortality' (1999: 8). Thus, metaphor is also a structural instability in the archive built between two dancing bodies as each reads the other, a very foundational ambivalence in the asymptotic intimate relation of language to body, of body to body, of one to another, always approaching, never reaching, in Jorie Graham's formulation: '[the] body an arrival/you know is false but can't outrun' (1980, 38–9). Metaphor insures 'a relation of indirection' (Irigaray, 1996: 109) in 'I love to you' and dance to you; it is reminder, a 'sign of non-immediacy, of mediation' (109) between self and body, body and body; like 'to,' it is 'the site of non-reduction of person to object' (Irigaray, 1996: 110).

Metaphor organizes intimacy in dancing communities by inserting play into the strategic protocols of technique. But this is not unmitigated free play. Metaphor is also subject to the same disciplinary regimes as other social technologies of border intimacy. Though my Pilates sessions over the years of working with Chip, Xavier and Jeanne are intimate transactions, they are also commercial ones; class fees are paid and propriety is called for.

Consider propriety an ensemble of 'minuscule repressions' (Mayol, 1998b: 17) and performances 'through the presentation of the body [by] a manipulation of social distance . . . expressed under the negative form of a "how far is not going too far?"' (21). Propriety, like 'to' in 'I love to you,' is also a 'guarantor of indirection (Irigaray, 1996: 109); it is a reminder that the technology of figural intimacy, metaphor, regulates the body that touches us, reads us, even as it makes that body safely available to be read. As Mayol observes, propriety plays the role of reality principle which, when invoked, allows one to seize corporeal verity and metaphoric ambiguity, and deploy both to open and to foreclose. It enables the diverse membership of dancing communities to submit to the intimacy of being read, then close the literal and metaphoric dressing-room curtains to avoid or manage it. Propriety is an escape clause in this intimate sociality. No surprise that Mayol finds it 'particularly pertinent' to intersections of 'the consumption and appearance of bodies' (1998b: 19). At its most banal, propriety offers an alibi to safely contain intimacies in Pilates. It transforms intimacy back into the sociality of consumer and service provider where any attendant relational and affective labor is simply the 'unaccountable surplus in the strict logic of the exchange of . . . services . . . the fruit of a long, reciprocal habituation in which each person knows what he or she can ask of or give to the other in hopes of an improvement of the relationship to the objects of exchange' (Mayol, 1998b: 20).

And yet . . . When I consider the touch and talk, the histories and fantasies at play in the Pilates studio – even against eminently more accessible narratives of commerce, or discipline – it is the generative ingenuity of diverse intimate, embodied relations I also see, the extraordinary inventiveness of this proprietary and metaphoric lovers' discourse I also hear. Perhaps this is why such a heterogeneous group of clients returns, week after week and year after year. Foster(1996), drawing on very different practices and using very different terms, suggests that, in this kind of intimacy, this resourcefulness, lies a corporeal politics of compassion, and I agree. Such a communal politics begins in the intimacy of complex, sustained arrangements like those in the Pilates studio where instructors and clients have the permission, indeed a practical and relational obligation, to physically read and move with another body, even as they are also read and moved.

PART III   ROMANCING MONSTERS, CONSUMING LOOKS

> *When ideals turn out to fail as objects of belief and models of legit-*
> *imation, the demands of cathexis are not disarmed; they take as their*
> *object the manner of representing those ideals.*
>
> (J. F. Lyotard, *Postmodern Fables*, 1997)

Dancing communities are forged through material, interpersonal exchanges like those in the Pilates studio. But they can also emerge through a deep, if vicarious, sense of ownership of, or communion with, a performer or performance. This is intimacy of another kind in another place: one of looking, imagining, consuming, longing, from the position of the audience. This intimacy may involve subsuming differences, or highlighting them, in operations that are imperialistic, exoticizing, fetishizing, and radically humbling at the same time. Yet this intimacy, this spectatorial politics of sociality, is also organized and enabled by technique. As in the Pilates studio, it arises from seemingly casual arrangements – attending performances – that deepen over time and sustained contact to become something more.

Possessing performance and/as the play of the vicarious

Technique's protocols of legibility aside, rendering dance discursive is often a fraught process for individuals and communities. As Foster notes in her introduction to *Choreographing History*[7]: '*How to write a history of this bodily writing, this body we can only know through its writing. How to discover what it has done and then describe its actions in words.* **Impossible. Too wild, too chaotic, too insignificant. Vanished, disap-peared, evaporated into thinnest air, the body's habits and idiosyn-crasies, even the practices that codify and regiment it, leave only the most disparate traces**' (1995: 4; italics and boldface in original). But this is not only a matter of navigating the transition from 'the moved to the written' (Foster, 1995: 9). There are also the demands of cathexis, the persistent, affective excess of performance pressing to be accounted for, described and, some would say, resisted. Peggy Phelan, for example, argues that this emotional afterlife,

> [t]his transformative becoming is the almost always elegiac function of performance theory and writing, if not performance itself. Our admiration for performance tempts us beyond our reason to make it ours, for better or worse. The challenge before us is to learn to

love the thing we've lost without assimilating it so thoroughly that it becomes us rather than remaining itself.

What lies before the field of performance studies is precisely a discipline: a refusal to indulge the killing possessiveness too often bred in admiration and love. The lessons we most need to learn are lessons in mourning without killing, loving without taking.   (1998: 11)

Phelan is writing about the inevitable trajectory of representation toward disappearance, but the performance, and the communal, intra- and interpersonal investments it may inspire, don't end when the show's over any more than the affective investments organized by dance technique completely dissolve outside the dance studio at the end of class. Her invocations of elegy, of mourning and loss, recall Jonathan Dollimore's assertion that, in 'Western culture the "tragic vision" has been one of the most powerful means of containing and sublimating desire' (in Sippl, 1994a: 23). Why, when beckoned by the emotional, transformative invitations to 'save' performance, *must* we deploy discipline to resist temptation? Why turn our backs on the longing to make performance ours? Why must we abandon the desire to envelop performance and hold it to us, or the desire for it to envelop us? Such a move would disavow one powerful component of performance's social, cohesive force. After all, exactly where are the boundaries between 'us' and 'performance itself' for members of dancing communities? Can't the impulse to hold onto performance, however futile, be recast as a move to a deeper, sustained and socially productive intimacy in/as community? (See also Dolan, 2005: esp. pp. 168–9.).

Irigaray sounds a note similar to Phelan's as she argues for an interpersonal ethic of love in language that touches and touches upon: 'The *touching upon* cannot be appropriation, capture, seduction - to me, toward me, in me - nor envelopment' (1996: 125). But Soyini Madison offers a counter-view, one in which our embrace of performance is far less hygienic and abstemious. In this view, desire, seduction, and longing will out, indeed *must* be acknowledged and even indulged if we are to theorize the interpersonal and communal power of performance. Here:

performance *performs* like a promiscuous lover. Enticing this one and that one, wanting everyone to love her and wanting to love everyone in return. She seeks and she is sought. . . . Indiscriminate in the pleasure she gets and gives, there is a natural striving for her, and through all her encounters and travels, there are those who speak of her very well and there are others who live cautiously inside her presence. I know some who do both. Now, I will admit, within my own metaphor, I am guilty of gluttony. I have taken Performance for myself (sometimes she was given to me with more love than I deserved). Like a possessive devotee (or a lover) I want to hold on to her, not wanting her to be with anyone but me and my kind; because we've known her for a longer time – who she is and how to treat her.

(Madison, 1999: 108)

As Eric Doxtader argues in his reading of Irigaray's *I Love To You*: 'In a culture that abstains from speaking desire's speech, we . . . censor experience. We leave ourselves without the means to question, without the courage to express what turns us on to those we care about. Worst of all, we implicitly perform the notion that desire is not the subject of communication. . . . words of beauty, intimacy and lust are repressed by the cult of abstinence (2001: 216–17).'[8] To desire to save performance and preserve it as a tool for interpersonal and communal intimacy is not necessarily to exhaust, betray or kill it. It is, however, to communicate, even if this communication is a conversation of looks[9] even if it is an intrapersonal or communal conversation wherein, fantastically, we engage both the process of loving a performer we recall and recreate, and who we become in this process (Sippl, 1994a: 41).

Suppose we indulge in some tactical possessiveness, less to hold fast to performance, perhaps, than to revel in and to query how performance possesses us individually and collectively, within and across taxonomies of difference? What new social and interpersonal possibilities are opened up by saying yes rather than no to possessing performance? Barthes has written: 'Political liberation of sexuality: this is a double transgression, of politics by the sexual and conversely. But this is nothing at all: let us now imagine reintroducing into the politico-sexual field thus discovered, recognized, traversed and liberated . . . *a touch of sentimentality*: would that not be the ultimate transgression? the transgression of transgression itself? For, after all, that would be *love*: which would return: but in another place' (in Pollock, 2001: 203). I want to argue for the social force and utility of a possessive love of performance, one characterized by, indeed demanding, giving and taking, seduction, longing and assimilation – love in another place of spectatorship – the realm of the vicarious.

I am guided by Diane Sippl's provocative assertion that we have foreclosed a generative field of play and, in so doing, the potentials of performance and spectatorship, by ignoring the 'menacing and comforting' power of the vicarious (1994a: 23).[10] She writes: 'Discovering and valuing the prospect of "eroticizing the social" (as opposed to merely "liberating the sexual") in our relations with each other, we may also come to appreciate different ways of looking – [at performance], at the world, and at the interpersonal screens [and stages] we construct' (Sippl, 1994a: 23). Virtuoso performance, with its evocative excess and its seductive calls for cathexis, for emotional engagement and investment, offers an especially fertile field for interpersonal, vicarious, intimate conversations between artist and spectator/critic. These are conversations enabled by dance's technical protocols for organizing bodies, even when bodies seem to evade or transcend them. Such conversations offer productive tools for theorizing solidarity, even intimacy and friendship, in spectatorship, and for reconceiving spectatorship as a form of social cohesion in real and imagined communities.

Faced with the virtuoso performer, Phelan's and Irigaray's ethics of relationality and resignation fail me utterly; it seems impossible to maintain a discourse of indirection *vis à vis* the virtuoso, even as this indirection proclaims its own inevitability. It isn't that Phelan's and Irigaray's insights are untrue here – it's that they seem constrained and evasive, particularly when compared to the social possibilities offered by demanding and compelling performances. In her chapter 'You Who Will Never Be Mine,' Irigaray seems to approach my dilemma: '. . . I draw myself to a halt before you as before something insurmountable, a mystery, a freedom that will never be mine, a subjectivity [and a physicality] that will never be mine, a mine that will never be mine' (1996: 104). Yet she does not capture what, for me, is the next crucial turn; as Madison (1999) indicates, this is possessive devotion. Somehow, if only in gaze, if only in language that touches upon, if only vicariously, this virtuoso 'you' of the performer MUST be mine.

Here, I take up Jill Dolan's challenge, 'How can we chronicle an audience's response, in the moment of performance?' (2005: 169). My goal is to examine the construction of intimacy in performance through looking and longing, operations at the intersection of the interpersonal and social aspects of spectatorship. The virtuoso performer is an especially evocative site for observing this process as s/he makes special demands on the viewer/critic. I am trying to describe what this exchange is like, to construct some metaphoric technology to organize this relationship of dancing and observing bodies for, as Foster observes, 'metaphors, enunciated in speech or in movement that allude to [the body] give [it] the most tangible substance it has' (1995: 4).

Defining virtuosity, engaging monstrosity

In some significant ways, all virtuosi are alike. As David Palmer observes in his discussion of Paganini, an expansive, generic definition of virtuosity 'posit[s] it as the *art* of incredible skill which displays a heightened sense of self-expression, evokes a distinctive affecting presence and transforms ways of viewing human agency' (1998: 345). While he characterizes virtuosity as a property of individuals, Palmer also acknowledges that '[t]he pervasive value of the virtuoso's performance resides in its allotted capacity to translate the community's ideals concerning the display (or yield) of extraordinary talent' (1998: 345).[11] 'Talent,' he argues, is 'a transformative and illuminating feature of a work' (344). In this view, virtuosi deploy talent in community-endorsed, heroic reimaginings of individual and human agency. Other dimensions of 'difference' are subsumed in, and contained by, the 'extraordinary' nature of the virtuoso's 'talent.' S/he stands apart from conventional taxonomies of difference, but is enfolded into community on the basis of an 'extraordinary' difference.

Yet such a view underplays the social, affective work which brings virtuosity into being, labors that exceed any 'self' expression of the artist, but instead rebound between performer and spectator in a vicarious relation that undermines agency even as it seemingly celebrates it through fantasies of extraordinary difference. It is this difference and the excesses of pleasure and anxiety it inspires above and beyond 'ideals,' that is the currency for the relationship I will call virtuosity.

I characterize the commonalities of virtuoso performers differently. In the largest sense, virtuosity is one of the 'typical "plots" which histories of bodies are destined to inhabit' (White, 1995: 230). In cases of exceptional performing bodies, these plots include heroism, mastery, talent: vocabularies used to characterize corporeal difference in reassuring terms of the autonomous artist and the communities that produce and endorse them. The legitimation by which such a plot secures currency is its own self-evidence, in a manner both authoritative and precarious; you 'know' virtuosity 'when you see it.' Thus, virtuosity as 'typical plot' seems simultaneously very empty and very full of representational and explanatory potential.

How is such a plot inhabited, set into motion? What kinds of affective work does it do? These questions call for a more intimate view of virtuosity. I suggest that, within the 'typical plot' shaping the extraordinary performer/performance, are conversations which position the dancing body and the observing body in technical protocols, in language and in desire. In so doing, the artist and the spectator create a social, inter-pretive occasion. Virtuosity, as I use it here, is not a state or quality of individual bodies. It is a very specific technology of, and occasion for, both a corporeal and an affective relationship. Here, as in all dancing communities, this relationship between two differently framed bodies is organized by technique and spectatorship, then inserted into language. At every turn, this process is, in Mark Franko's terms, 'glued to the expectations, delusions, and agendas' deployed by artists and audience (1995: xii).

Virtuosity is a communal organizing fiction. It contains a social story of looking, of engaging corporeal and contextual difference, not through invocations of the essential and the autonomous, but rather through the power of the vicarious and, specifically, through projections onto the screen of the exceptional performer. Projection is the ongoing, affective identification we observers have with another body, enmeshed with our desires, fraught with the fetishes and fantasies by which we make our own identities: we as both observing and dancing bodies project to construct ourselves (Sippl, 1994a: 32). And there is pleasure, valid-ation and terror in being the critical, observing body, the interlocutor who co-creates this conversation of virtuosity. Indeed, I would argue, using Susan Stewart's formulation, that virtuosity is both a communal, conversational 'structure of desire,' and a structure of anxiety that 'invent[s] and distance[s]' interlocutors (1993: ix). Palmer characterizes virtuosity as a rhetorical artifact of Romanticism, and this legacy persists in contemporary, uncritical uses of the term. I want to reframe the relationality of virtuosity, not as the resuscitation of Romantic agency, but as one example of the unsteady mutualities of intimacy in dancing communities.

If, as Foster asserts, metaphors substantiate the body, how do we metaphorically stabilize the body of the virtuoso? What image of the extraordinary body is constructed as the spectator remakes herself, and the performer, in this projective play? Perhaps, in the words of another virtuoso, the late Yehudi Menuhin, spectatorship makes this extraordinary corporeal screen into a 'sacred monster' (*60 Minutes*, 3/21/99). Consider 'sacred' in Bataille's interrelational sense, where the 'sacred object and subject interpenetrate or exclude each other but always, whether in association or in opposition, complete each other' (1984: 115).

'Monsters are meaning machines,' writes Judith Halberstam in her survey of gothic monstrosity *Skin Shows* (1995: 23). Moreover, they are meaning machines intimately connected to excess, to the desires

and anxieties of performance and spectatorship, even etymologic-
ally. As Brian Noble suggests, the very word 'monster' is linked to
'demonstration,' to 'the showing of visible evidence' (in Mitchell, 1998:
65). Because of its spectacular visibility, the monster functions 'as a
dialectical Other or third-term supplement . . . as an incorporation of
the Outside, the Beyond – of all those loci that are rhetorically placed
as distant . . . but [may] originate Within' (Jeffrey Cohen, 1996: 7); it
embodies the affective excesses of projective, vicarious relations. Cohen
writes, 'Any kind of alterity can be inscribed across [or constructed
through] the monstrous body' (1996: 7). The alterity of virtuosity
is one compelling example. Both the virtuoso and the monster
'symbolize alterity and difference in extremis. They manifest the
plasticity of the imagination . . .' (Weiss, 2004: 125). Like the virtuoso
performer, monsters are 'remarkably mobile, permeable, and infinitely
interpretable bodies' (Weiss, 2004: 21).

There are monsters and there are monsters; the virtuoso as monster
both participates in and extends traditional taxonomies of the category.
In her survey of nineteenth-century fiction and twentieth-century film,
Halberstam suggests that Frankensteins and Freddie Krugers 'demarcate
the bonds that hold together' systems of relations 'to call attention
to borders that cannot and must not be crossed' (1995: 13). Vampires
and werewolves 'reveal that difference is arbitrary and potentially free-
floating, mutable rather than essential,' threatening 'to destroy not
just individual[s] . . . but the very cultural apparatus through which
individuality is constructed and allowed' (Halberstam, 1995: 12). But
the sacred monster 'polices the borders of the possible' (J. Cohen,
1996: 12) differently, not through interdiction or critique, but through
extraordinary spectacles of virtuous discipline in performance. In doing
so, the virtuoso as sacred monster rewrites plots of possibility for other
bodies generally, even while demonstrating specific bodies', including
the spectator's, inability to execute this virtuous discipline themselves.
Thus, virtuoso as a social position founds community both in spite
of difference ('we are all witnessing the demonstration together') and
because of it (it's extraordinary nature, which we cannot, ourselves,
enact).

While the general contours of virtuosity are the same across media,
every sacred monster is unique; every technique organizes its own
monstrosity, and every community engages its virtuoso monsters on
its own terms. As Weiss observes, 'The logic of monsters is one of
particulars, not essences. Each monster exists in a class by itself' (2004:
124). Los Angeles-based Japanese butoh artist Oguri, and the butoh

technique he deploys with such virtuosity, is my exemplar in the discussion that follows. Oguri's work, and the technical protocols of butoh, are especially amenable to an examination of the border intimacies and communal possibilities of spectatorship. Jeffrey Cohen observes that one constitutive element of monsters, sacred and otherwise, is that they always escape (1996: 4). I was prompted to rethink virtuosity because Oguri repeatedly, discursively escapes from me.

Oguri: sacred monster

I first saw Oguri[12] perform at Los Angeles Contemporary Exhibitions (LACE) in May 1991, his local debut. He had performed in Japan, his birthplace for many years; after working there in the visual arts in his early twenties, he met Tatsumi Hijikata, the founder of butoh, who changed his life. Hijikata recommended that Oguri study with Min Tanaka, another butoh pioneer, who had established a farm/commune/training facility for dancers and others interested in his technique. Oguri spent five years on the farm, dancing and doing manual, agricultural labor. There he met his future wife, Los Angeles dancer Roxanne Steinberg.

Oguri rewrote the possibilities of the butoh body, indeed the dancing body, for me in this early performance, but I was familiar with his idiom before this. In the summer of 1988, my first in Los Angeles, I took classes in Body Weather movement, also at LACE, from dancers Melinda Ring and Roxanne Steinberg; Roxanne would marry Oguri two years later and they would settle in Venice, California. I was compelled by the technique. 'Body Weather' is a formulation of Min Tanaka's. As Melinda, Oguri and Roxanne characterized it: 'Body Weather Laboratory is an open opportunity for anybody determined to be involved in physical expression. A demanding attitude towards a thorough reexamination of the body is one of the main themes of BWL work' (Flyer, 1993). Min Tanaka likewise suggests that Body Weather movements interrogate corporeality, and that this interrogation is 'not only for performers, it is for anybody. School teachers, housekeepers, monks. Everybody. We have 16 [Body Weather] stations in Japan and other parts of the world' (in Stein, 1986: 148).

Body Weather calls for demanding physicality as well as a demanding attitude. Sessions include rigorous moves across the floor, often in low, quadricep-crunching crouches, improvisations, and my favorite component: manipulations in which partners massage, stretch and flex each other's limbs. The movement vocabulary draws loosely from butoh, originally *ankoku buyo* (Kurihara, 12) or 'dance of utter darkness,' a modern expressionist Japanese dance form founded by the late Tatsumi Hijikata in 1959. Shaped explicitly as provocation, butoh is best known for its bowlegged crouch with weight to the outsides of the feet (*ganimata*), its severely attenuated gestures, and its exaggerated, grotesque facial expressions and bodily convulsions (*beshimi kata*). Different butoh practitioners inhabit the technique differently, with some emphasizing lyricism, like Kazuo Ohno, and others a darker, more nihilistic and erratic vocabulary, like Min Tanaka, mentor to Oguri, Roxanne, and Melinda. All three had trained at Tanaka's farm, though Roxanne and Melinda spent less time there than did Oguri.

Eventually, Roxanne, Melinda and Oguri went their separate ways, in part because of Oguri's commitment to forming a company of dancers from the group who regularly attended the trio's Body Weather training, a commitment Melinda did not share. Founded in 1993, core members of Renzoku ('Continuum') included Boaz Barkan, Jamie Burris and Dona Leonard, as well as Roxanne. The company was originally housed at La Boca/The Sunshine Mission in Los Angeles; they performed regularly in a wide range of venues, both in Los Angeles and internationally. Members continue to perform with Oguri in his solo work.

Roxanne told me that she and Oguri dance out of what she called a strange obligation because, she said, they took it seriously when a member of their audience told them she never came in with a problem their dances couldn't solve. This sense of obligation, of bond between performers and spectators, is one element sustaining the extended and diverse audience-community that follows Roxanne and Oguri's work, and has done so for over a decade. The community includes Body Weather collaborators like Anglo-American dancer Jamie Burris, Israeli artist Boaz Barkin, and Asian-American Sherwood Chen who, after meeting Oguri, took a leave from college to 'follow the rice' in Japan, as well as assorted members of the entertainment industry and performance art scene in Los Angeles, homeless women at LA's La Boca/The Sunshine Mission, and a very attentive Chinese Crested dog who showed up at performances, emerging from the gym bag of her leather-clad owner. Indeed, literal invitations to the community to 'join the dance' are prominent features of their work, whether teaching classes at La Boca, or at the Electric Lodge in Venice, or on performance 'caravans' like those discussed below. I found Roxanne's words ironic because I never come in with a critical solution that these dances couldn't make problematic.

The issue is this: Roxanne is a dancer of enormous grace and power but Oguri overwhelms. His body, his movement vocabulary, his ethos are almost excessively present. He is always almost too much there and not, or not only, because of the choreography or *mise-en-scène*. His work is beautiful, terrifying, 'at the limits of the possible' (Barthes, 1985: 286). His excess of presence, and my joy and anxiety in the face of it, leave me searching for language up to the task of representing not only the dance, but how I am remade through it. And in the search for accurate language, I lose him.

Perhaps there is a straightforward material, contextual explanation for the exceptional difference, the excess, of Oguri's performance. Perhaps his genealogy accounts for this: five years spent on Min Tanaka's farm as opposed to fewer or none for other dancers. Perhaps this difference is constructed through the semiotics of choreography: he is positioned to be read as exceptional. Perhaps an empirical, close reading of his technique will help me explain; maybe his execution of butoh's protocols for reading the body are more intense and sustained, or less predictable. Or perhaps it is not Oguri's body that is lost at all in these accounts, but my own body, observing and consuming with a spectacularly hungry orientalism. Perhaps he fits what 'exceptional' or 'exotic' or 'excessively present' looks like for me and for others in his audience-community.

These are all so, but something is still missing. The sum of these explanations does not suffice when accounting for the affective connection between my active spectatorship and Oguri's excessive dancing body. Local reviewers resuscitate typical plots to characterize his difference; he is, they say, 'a master,' a 'true artist,' and, in terms that span the culture/nature binary, both a 'specialist,' and 'pure instinct' (Looseleaf, 2002; Rauzi, 1998; Segal, 1998). But these are insufficiently attuned to specifics of Oguri's extraordinary movement, to how these specifics activate and evacuate the terms used to characterize them, and to how this dance births and engages structures of anxiety and desire through the powerful, projective play of the vicarious. Only a deeply relational, affective account of virtuosity, infused with the wonder and terror of the monstrous details, can account for the social force of this work for his audience.

My readings of Oguri's work span 11 years and are drawn from both live and videotaped performances. The virtuoso performance is replete with the aura of 'liveness'; Palmer (1998) asserts, for example, that witnessing the spectacle of a Paganini concert could generate the potential for *communitas*. Indeed the marketing of virtuosi relies heavily on this very notion of witnessing, perhaps verifying, the presence of extraordinary performers. Philip Auslander, on the other hand, is deeply suspicious of this aura, arguing:

> I quickly become impatient with what I consider to be traditional, unreflective assumptions that fail to get much further in their attempts to explicate the value of 'liveness' than invoking clichés and mystifications like 'the magic of live theatre,' the 'energy' that supposedly exists between performers and spectators in a live event, and the 'community' that live performance is often said to create among performers and spectators. In time, I came to see that concepts such as these do have value for performers and partisans of live performance.... But where these concepts are used to describe the relationship between live performance and its present mediatized environment, they yield a reductive binary opposition of the live and the mediatized.   (1999: 2–3)

Rather than reinvigorate this live/mediated binarism, Oguri has it both ways. He performs live and also works creatively with video, frequently collaborating with his sister-in-law, Morleigh Steinberg. Indeed, he actively uses video to link small subsets of his dancing community to the whole across space and time. His 'caravans,' discussed in the following pages, are especially illustrative. Even in live performance, his staging often suggests the cinematic (Figure 1.2). Yet, whatever the context, his virtuosic invitations beckon. A closer look at the play of the vicarious in these invitations allows us to trouble the distinctions between live and mediated performance, and ask what kinds of relational opportunities and permissions are offered to spectators in each mode.[13]

Jill Dolan observes that it is 'through technique and precision that presence gains power' (2001b: 470). Technique and precision likewise shape the power of the mediated body. Oguri's videographers take special care to preserve the attenuated quality of many of his movements and the deliberation with which they unfold over time. Editing in these pieces does not seem to impose an arbitrary pace on his dances, but instead generally presents them in self-contained sequences that parallel the spectators' experiences of his performances. While camera positions shift, this is also not appreciably different from performances like 'TOE/Earthbeat 2001' (May, 2001) where both Oguri and the live audience moved around and through the performance space, here the Electric Lodge in Venice, California. Likewise, stationary cameras shooting him head on in long takes recall pieces in which Oguri demanded fixed audience points of view, as in 'Onami/Menami' (May, 1995) where he

Figure 1.2 Oguri, *Effect of Salt*. California State University, Los Angeles State Playhouse, 1996. Los Angeles, California, USA. Photographer: Roger Burns

and Renzoku performed outside of the loading dock at Highways in Santa Monica, CA, framed by the rolling metal dock door.

Technologies of monstrosity

The protocols of butoh technique offer 'pre-existing patterns of fascination' (Lebeau, 1991: 256) that work across both Oguri's live and mediated performances to invite the spectator into a projective, vicarious relationship. If every technique constructs its own monstrosity, butoh does this more explicitly than most. The first 'structure of fascination' Oguri deploys across both stage and screen is the extreme grotesquery of the technique itself. This is not his own innovation, even if his execution of it is exemplary. Images of horror have been intrinsic to butoh from its inception (see, for example, the writings of founder Hijikata, especially 'Plucking', 2000a). As noted above, the face and body of the dancer, including fingers and toes, are contorted; in Oguri's case, these contortions are extreme. His joints seem to have migrated out of their sockets to generate new kinds of angles, new technologies of leverage. His torso is cantilevered twenty degrees back from his pelvis, or folded 30 degrees forward. His head and shoulders loll back and stay there, generating an uncanny 'other face' as his lips become a browline and his creased forehead another mouth.

In an interview I conducted with Oguri and Roxanne in their home, he wondered aloud about why more people don't laugh during his performances. Can't they see that sometimes these images are so funny? Well, I suggested, perhaps the aura of *ankoku butoh* is too powerful. After all, it is often translated as 'the dance of utter darkness' and his movements are, frankly, unsettling, even monstrous as they strain corporeal possibility. He appears in agony, or in ecstasy, or beyond either. Oguri smiled. 'Lovely darkness,' he said, with a touch of the trickster. 'Lovely monsters' (Figure 1.3).

Grotesquery is an interpretive invitation, and a particularly complex one. As Geoffrey Harpham observes, 'in the case of both metaphor and the grotesque, the form itself resists the interpretation that it necessitates' (in Susan Klein, 1988: 29). I would put this another way. Grotesquery, and particularly the monstrous body in performance, exposes interpretation as asymptotic, as inevitable and futile, inexhaustible, yet always, of necessity, in motion. To read the grotesque, virtuoso body is to both approach it and recede in fascination and unease. Writing specifically about butoh, Klein suggests that this interpretive ambivalence has a parallel in spectators' affective ambivalence: 'I would go on to say that although we feel repulsed, there is at the same time a reluctant sense of identification: the grotesque, like the [Freudian] uncanny, "is nothing else than a hidden familiar thing that has undergone repressing and then emerged from it."'(1988: 29). But in virtuosity and the vicarious romance it inspires, there is, I argue, relatively little reluctance. Instead, there is the desire to hold fast to, even to inhabit, the exceptional body, particularly in light of protocols of reading butoh that reveal grotesquery as expertly crafted. This is true even when such a

Figure 1.3 Oguri, *Behind Eyeball*. La Boca/The Sunshine Mission, 1995. Los Angeles, California, USA. Photographer: Kevin Kerslake

hold is clearly impossible. Here the exceptional performer's uncanniness resides in my capacity to remake him, in Schechner's and Winnicott's formulation, as 'not me' and 'not-not-me,' as both the craftsman and the monster exceeding and inspiring interpretation, with the possibility of disavowing either or both (Schechner, 1985: 109–11).

Klein suggests that the ambivalences of butoh technique block critical responses to the work and allow audiences access on a 'gut' level. But, in the case of a virtuoso like Oguri, the critical and visceral are not opposed. His grotesquery is an interpretive occasion that never exhausts itself, always generating and regenerating new accounts and new possibilities for spectators to inhabit. This critical effort has its counterpoint in the performer's precise and effortful execution of grotesquery. As Bonnie Sue Stein observes, 'Because butoh is so obviously demanding, spectators who may not like it – who may even feel uncomfortable confronting such intensity – still respect the experimentation and the performance skills required' (1986: 112). The invitation to the dancing community involves reading what Oguri does and why, as well as how he does it, without the mystifications of 'genius' or 'pure energy.'

Both on tape and in the theatre, Oguri's efforts are visibly, relentlessly marked: by beads of sweat on his brow, sweat stains on his cloths, redness where limbs hit the ground, dirt building up on the soles of the feet. In live performance there is the addition of the sound of his breathing, but Oguri's heaving chest in video performances also signifies this effortful grotesquery.

Though the vicarious invitation generated by these labors in service of the grotesque operate across Oguri's live and taped performances, they inflect spectatorship in each medium differently. In live performance, Herbert Blau reminds us, we are continually confronted by 'the fact that he who is performing can die there in front of your eyes; is in fact doing so' (1982: 83). This confrontation is especially acute, I believe, when the images the performer embodies are both grotesque and clearly demanding; it generates a desire for identification with the agency of the exceptional body as well as anxieties about the fate of that body, and, vicariously, our own, within the performance. Perhaps the immediacy of this desire and anxiety accounts for the paucity of laughter at Oguri's live performances. As he sustains impossible poses, or jerks from one to another in spasmodic frenzy, the spectator asks, 'Can he make it? Could I make it?'

While Oguri's effortful grotesquery, and its attendant spectatorial desire and anxiety, also characterize his video performances, there is an additional dynamic at play. Cut loose from the demands of immediacy, spectators' identifications and projections have more freedom to roam. Fantasy does not have to keep up with the movement in real time and, in the endless half-lives of videotape, mortality is not so obviously in play. Thus, in a dark theatre, in a video documentary of his project *Height of the Sky*, Oguri, squinting against blowing dust and the sun's glare, opens one eye. Its whiteness, stark against his ocher pigmented skin and the brown monochrome of the landscape, looks like an cartoon eye in a Looney Tunes telescope. People laugh.

Oguri's taped performances facilitate heightened and more leisurely regard of key elements in his work. Further, they allow for vicarious participation when actual participation in productions is impossible. Consider *Height of the Sky*, mentioned above, as one example. Dances in *Height of the Sky* were developed and performed by Oguri on a series of caravans into Southern California's Joshua Tree National Park (Figure 1.4). There were three caravans between 2000 and 2001; participants included Roxanne and Morleigh Steinberg, videographer Charlie Steiner, photographer Roger Burns, and Body Weather members Jamie Burris and Sherwood Chen. Invitations to join the caravans were widely disseminated, but vicarious participation was also an explicit element of the piece: 'Can an audience journey with the dancer not simply as spectators but as travelers – turning into the stream, walking with the

Figure 1.4 Oguri, 'Draught,' *Height of the Sky*, 2002. Joshua Tree National Park, USA. Photographer: Roger Burns

dancer, creating time?' (promotional material, 9/24/01). Still, I argue, these differences in the literal places of spectatorship, and in repeatability, between Oguri's live and mediated performances are differences in the degree, not in the kind of vicarious communal participation his work invites.

Other patterns of fascination ensure that these invitations are equally, if somewhat differently, compelling across media. For example, the grotesquery of butoh technique is intimately linked to the exotic otherness of the dancer. Certainly, part of this exoticism is the orientalizing of difference. Though, as noted below, Oguri's work often complicates ethnic identification in performance, he is, in many of his pieces, clearly readable as a Japanese man. This is emphasized when he is dancing with white performers like Barkan, Burris and Steinberg. Vicarious miscegenation is one spectatorial possibility undergirding the longing to possess this virtuoso.

Barthes characterizes 'the Orient' as 'the possibility of difference' (1982: 3), and this possibility is deployed, both by Western and Japanese critics to characterize butoh. Consider these observations from Bonnie Sue Stein and Kazuko Kuniyoshi. Stein writes: 'The work of these Japanese artists is so thorough and so "Japanese" that Westerners sense a searing honesty' (1986: 112). Kuniyoshi adds: 'Western theatre and dance has not reached beyond technique and expression as a means of communication. The cosmic elements of butoh . . . are welcomed by Western artists because they are forced to use their imaginations when confronted by mystery. Butoh acts as a kind of code to something deeper, something beyond themselves' (in Stein, 1986: 114). Contrary to such metaphysical and essentialist generalizations, reading exotic difference in butoh is actually fairly complex. First, the 'Japaneseness' of butoh is itself ambivalent. Klein asserts that the form was deployed by the Japanese avant-garde to resist Western cultural imperialism (1988:10–14), but it was never widely viewed by large local audiences. Both Hijikata and Tanaka were thrown out of the Japanese Modern Dance Association, the former in 1959, the latter in 1972. Further, with notable exceptions like Hijikata, who never left Japan, butoh was a *gyaku-yunyu*, a 'go out and come back' phenomenon: largely an export. Key performers like Tanaka and Kazuo Ohno, and companies like Dai Rakuda-kan, toured extensively. Sankai Juku, arguably the most 'commercial' butoh group, is based in Paris and gave its early performances in Europe (see Stein, 1986: 114; Klein, 1988: 19).

Further, as I have argued elsewhere (1990), butoh's deployments of assaultive, grotesque imagery in the West should be read against Western performances which were both contemporary and shared the same contexts of reception, including the same audiences and venues. As Kuniyoshi's authoritative 'Butoh Chronology' (1986) indicates, butoh shares the Western cultural stage with happenings, fluxus performance, and performance art, forms that also take on and critique Western aesthetics. On tour, butoh performers in the Unites States danced in performance art spaces and events: Los Angeles' Olympic Arts Festival, Chicago's Mo Ming Center, and New York's La Mama to take only three examples. Their local contemporaries were working in styles and idioms which were also improvisatory, spasmodic and random. The exoticism of Japanese butoh performers might have centered on 'authentic,' as opposed to 'local' artists, but the general aesthetic presumptions deployed by these bodies would be seen as part of an avant-garde, expressionist continuum with which many dancing communities would be familiar.

This 'authenticity' could itself be further complicated; in the case of nonagenarian Kazuo Ohno, it is the spectacular visibility of his age and seeming frailty, alongside his ethnicity and technical virtuosity, that structure and compel Western audiences' fascination. Oguri, and butoh generally, insert what Randy Martin calls 'the unruly polyvocality of difference' (1998: 123) into dance on many historical and

contemporary, literal and metaphoric stages. Thus, while exotic difference is certainly part of spectators' vicarious romance with Oguri
across media, the dynamics of reading this difference is more freighted
than simple ethnic essentialism and a metaphysics of 'the East' might
suggest.

Oguri's 'explicit body' in performance provides another structure of
fascination across multiple media. Rebecca Schneider characterizes the
explicit body as one that generates binary terror by imploding the
conventional dualisms organizing identity. As I discuss below, Oguri
certainly problematizes such structures in his work. Here, however, I am
using the notion of the explicit body in a more general way. In this
sense, Oguri's explicit body itself, apart from choreography, becomes a
screen on which the social and affective possibilities of spectatorship
are put into play (Schneider, 1997: 20).[14] He appears whitened, or burn-
ished red, with body makeup and powder; in high heels, work boots or
barefoot; shaven and unshaven; crusted with dirt; in a business suit, a
trench coat, in safari khakis; naked expect for his penis tied in a snood,
or covered by a *fundoshi*, a loose, diaper-like garment. In these various
guises, he reveals and complicates spectatorial possibilities for the spectacular male body on stage.

In *The Male Dancer: Bodies, Spectacles, Sexualities*, Ramsay Burt observes:
'An assertion of the physicality of the masculine body that challenges
normative conventions . . . has the potential to present otherwise invisible aspects of male experiences of embodiment' (1995: 73). These kinds
of challenges were intrinsic to butoh from its inception. Consider Min
Tanaka's early performance (1972), which was blacked out shortly after
it began and resulted in his expulsion from the Japanese Modern Dance
Association. Tanaka was naked, 'with just a bandage to hide my penis':
'The very narrow stage had a black curtain, almost closed, only one
meter opened. Behind was completely blue lighting. I was dancing with
a plant between my legs. Then suddenly I opened my legs and the
plant fell, and I also fell, into the audience. Then I climbed up to the
stage again and people recognized my wrapped penis. Then the lights
went out [were cut off]' (1986: 145). As Burt notes, 'Conventions generally dictate that no spectator should be shown the male body as if he
were the object of a pleasurable gaze. This is because the spectator is
presumed to be male and his dominant male gaze a heterosexual one.
For a man to look in an erotic way at a man's body is to look in a
homosexual way and homosexuality is a threat to the continuity of
homosocial relations, the latter being an essential element in the ways
in which men work closely together in the interests of other men . . . '

(1995: 72). Butoh challenged this presumptive heteronormativity of the gaze from the very first in 'Kinjiki' ('Forbidden Colors'), performed in 1959 by Hijikata and Yoshito Ohno, son of Kazuo Ohno, and choreographed by Hijikata (see Stein, 1986: 115; Klein, 1988: 9; Kurihara, 2000: 18). While Oguri's performances lack the erotic flamboyance of Hijikata's work, or the gender bending of Kazuo Ohno's cross-dressed performances (see Franko, 1995: 100–97), they do unsettle conventional plots for masculinity on stage.

Simple physical exposure, in both live and mediated performances, is one way Oguri organizes fascination and rewrites typical plots for male dancers. As Hijikata's writings, including his citations of Bataille on nakedness indicate, the exposed body was important to the butoh aesthetic from the first (see 'To Prison', 2000: 45). Oguri's body is remarkable; every vertebra, every rib, every muscle is visible, readable. When he performs in body makeup, his sweat seems to bleed brown or white; he is literally streaked by his efforts, a visible index to his labors for spectators. When he flings himself to the ground, or against a wall, little clouds of rice powder surround him. When he wears clothes, they eventually cling to him like weighty, sagging skin. The effects are more than visually arresting; they generate visceral, corporeal empathy. These are vicarious invitations into shared embodiment, whether in live performance or on tape. After one of his intense, spasmodic solos breaks into more lyrical movement, I have seen audiences slip back in chairs and exhale, as if from their own exhaustion. As he repeatedly spins and stamps into the rocky desert ground on a tape of *Height of the Sky*, I strain to see if he is barefoot and try to imagine how his feet are bearing up as he shuffles through a dry salt lake bed.

Oguri does not use simple nakedness. Like Tanaka and other male butoh dancers, he frequently wears a genital sock. The effect is a strange one. The penis is curiously both on and off display – indeed 'on' because it is so visibly concealed and constrained. It is difficult to read this physicality simply as asserting, exposing, celebrating or retarding phallic power and potential.

Amelia Jones has observed, 'the dis/play of the penis, while exposing the male body to the harsh violence [sic] of the spectatorial gaze, often simultaneously takes place as a means of aligning this organ with the phallus of artistic authority' (1998: 113–14). While it may be possible to argue that Oguri's technical prowess generates such an alignment, close readings of his performances lead to a different conclusion. His work oscillates between celebrating the trappings of agency and evacuating them.

In many performances, like 'behind eyeball Fu-Ru-I(shift)#2,' (La Boca/Sunshine Mission, Los Angeles, 24–26 February 1995), he wore a black business suit or a trenchcoat while wobbling on high heels, rolling to the outsides of his feet in them or stomping slowly, firmly, repeatedly, as in a slow-motion tantrum. In *Height of the Sky*, he attempts to don the coat of a safari suit, imperial explorer garb, but he flails about, with the coat covering his head and upper body as if it is attacking him. In this same piece, he is completely covered with stones, as in a grave, in the desert – a setting Angelenos would recognize from local news as a favorite place to dispose of bodies. When he rises from the stones and toddles off, it is not a monumental resurrection, but some odd rising of the undead. Oguri's virtuosity is no pure celebration of masculine, or even human, agency. Rather, his spectacular skill exposes this agency's contingencies, sometimes its futility, its smallness.

To characterize virtuosity like Oguri's as a fantastic, spectatorial conversation with a monster, one in which I remake both him and myself in a vicarious romance, is to reckon with the category crises he invites and inspires. As Jeff Cohen observes, monsters mark category crises as a general rule, unsettling relations between what we see and the taxonomies into which we place it (1996: 6).

Among the other categories Oguri unsettles are conventional taxonomies of difference. His performance on video, *Height of the Sky*, offers an especially interesting example. Here Oguri explicitly frames the project in terms of his ethnic identity as 'a Japanese dancer in America' (promotional material, 9/24/01). Yet to watch him, particularly when he is face down or when the camera approaches him from behind, it is difficult to tell what his ethnicity actually is. He could be almost any man: a sunburned Anglo, a Latino, a light-skinned African American. Indeed, at times he constructs images that explicitly invite the viewer to evacuate, not only taxonomies for reading human diversity, but those of reading general humanity as well. In one utterly remarkable image from Caravan II in *Height of the Sky*, Oguri is next to a desert tortoise, which moves around him and seems to regard him with a kind of leisurely perplexity. Oguri is sitting on his knees, bent forward so that the crown of his head rests on the ground. His arms are tucked under his body. Slowly he seems to fold his head under his clavicle; his shoulder blades separate, and the contours of his muscles and bones transform him into a tortoise shell. As the turtle plods off down a dry desert wash, we look back to Oguri in a partial headstand, pubic bones up, feet flexed, like the brown trunk of a Joshua tree.

Halberstam writes that monsters 'can represent gender, race, nationality, class and sexuality' (1995: 21–22). Which of these are important to you in your work, I once asked Oguri. 'It is very important that I am small,' he said, again evading conventional categories of difference. 'That is so important. That I dance as a small person.' Can you explain, I asked. And he told me a story of one of the lovely monsters who, along with Hijikata and Tanaka, crystallized his vision as an artist:

> When I was very young, I was playing at the playground in a ditch of muddy water under the swings. I was wet and I had been left alone. Seized by aloneness, I ran home and saw my father working next to the shed in our yard. Because he was working so hard, my father's tank top was wet from sweat and was transparent. His back muscles were gleaming and shined from the inside. The scene penetrated my being. He was human and not human, so beautiful and almost ugly. I was terrified and I fell in love and I wanted to become that back and to become myself seeing it. So small and so large.
>
> (4/25/94; see also a printed version in rehearsal notes for 'A Flame in the Distance,' California Plaza, Los Angeles, Tuesday, 23 September 1997)

This love story, this terror and beauty of the laboring body of the father and its vicarious re-embodiment in Oguri's longing, runs through his repertoire and offers one turn in theorizing his relationship to typical plots for organizing bodies into discrete socialities. Oguri opens up and makes not only explicit, but spectacularly visible the relation between dance and labor, in his words so beautiful and almost ugly. Early in his career, on Tanaka's farm, this relation was literal; dancers studying with Tanaka performed in the evenings and tended vegetables, rice and chickens during the day to support the community. But Oguri's technique presents the relation between dance and labor in a more attenuated way.

The description that follows is a composite, drawn from performances of a persona Oguri employs in many pieces. Perhaps this persona is expressed most emblematically in Morleigh Steinberg's video *Traveling Light* (1994). Oguri's black suit is his uniform in this work. Against scores of industrial noise, breaking glass, jackhammers, traffic, he becomes the sometimes broken, often spastic, salaryman, raging or shattered, always laboring: the body in crisis, he says, anonymous but absolutely singular. Critics grasp for appropriate metaphors, all vaguely industrial. He is

'Chaplinesque,' they write, 'a Borofsky sculpture' (e.g. Looseleaf, 2002). Announced everywhere in these dances is an anxious celebration of a body visibly at its physical limits.

Consider how different this image of dancer as laboring monster is from the conventional hero story of virtuosity in dance. Butoh generally offers a battle with technique that is clearly a battle. Oguri is not the idealized body of ballet, where the conspiracy of dancer and spectator produces a story of a technique so hard it looks easy, a hero story pure and simple. Rather, Oguri's virtuosity offers a spectacle of a technique that is hard and looks it, producing a body fragmented, 'categorized by virtue of a kind of analytic [and spectatorial] dissection,' where each limb or gesture or expression is cut into successive sites of labor and never completely reassembled as, or sutured into, the smooth surface of mastery (de Certeau, 1997: 21). Generally, the labor of dance is exposed by those who visibly have to work at it, that is, those who are not master-heroes but struggling novices. Oguri presents a particularly monstrous kind of virtuoso body: a disturbing hybrid, a laboring master, a 'form between forms,' a 'mixed category' (J. Cohen, 1996: 6, 7).

Beyond this, the spectacle of the laboring virtuoso complicates fantasies of heroic agency in another way. Oguri's labor is the mirror opposite of his audiences' physical inadequacy in the face of such effort. However much I long to inhabit this exceptional body, whichever way I vicariously embark on the caravan or join the dance, I confront my own limits. Fantasies of my vicariously available, reassuring, autonomous agency are undone as the gulf between my physicality and Oguri's is reperformed again and again. The longing, however, persists . . .

This longing across a gulf of physical difference is glorious and terrifying, and is itself a form of interpersonal and social infrastructure. To engage in a vicarious relation with a virtuoso is to allow oneself to be swept away by an upswelling of affect, by the necessity and futility of translating this into language so as to share it with others, by one's own insurmountable distance from, and connection to, the performer who inspired it. The virtuoso offers, not the 'calm feeling of beauty' but the agitation and terror of 'privation: privation of others, terror of solitude; privation of language, terror of silence; privation of objects, terror of emptiness; privation of life, terror of death' (Lyotard, 1988: 99). This 'privation,' this 'ambivalent enjoyment,' characterizes Lyotard's notion of the sublime in avant-garde art; it reminds us of 'another kind of pleasure that is bound to a passion stronger than satisfaction' (99), love in another place.[15] The sublime, Lyotard argues, 'is kindled by the threat of nothing further happening' (99), of the end of, the loss of, the performance, the performer, and our investments in both. I would suggest that this terrifying pleasure of the sublime is one constitutive element of the romance with monsters I call virtuosity.

Like all virtuosi, Oguri exposes the ambivalence of the intra- and interpersonal conversations that constitute spectatorship, conversations in which what Wislawa Szymborska calls my 'lone, nonconvertible, unmetaphoric' observing body dances itself, remakes itself, however provisionally or inadequately, against the ever-capacitating screen of Oguri's monstrous body. Here Oguri and I dance and undance the very distinctness Irigaray valorizes, 'the condition for a possible respect for myself and for the other within our respective limits' (1996: 117). In the romance of virtuosity, limits are both abandoned and reincarnated. This is a conversation of looks and the play of the vicarious, and in these looks and this play I embrace Oguri, consume him, inhabit him, lose him. We are sutured together and cleaved apart: his visible body, my invisible text. It is a vicarious, urgent and impossible transmigration.

*   *   *

The sustained, deeply affective engagements with the virtuoso, or among clients and instructors in places like Le Studio Fitness, remind us of the importance of interpersonal conversations in the work of both performance and community-building. Through looking, talking, touching, over the course of months and years audiences, performers, consumers and teachers generate productive, even intimate connections out of seemingly ordinary ones. In *Utopia in Performance*, Jill Dolan writes: 'The desire to feel, to be touched, to feel my longing addressed, to share the complexity of hope in the presence of absence and know that those around me, too, are moved, keeps me returning to the theater . . . ' (2005: 20) and, I would add, keeps clients returning to the Pilates studio. In the theater and in classrooms, technique organizes and shapes friendships by offering common languages, albeit those that can be diverted to other ends.

Amin and Thrift observe that even increasing surveillance cannot eliminate one crucial element of city life: novelty, the 'unexpected juxtapositions at all kinds of levels' (2002: 40). Dancing communities in the global city rely on the most abstract and the most concrete of these juxtapositions: the rigors of technique over against its fantastic permissions, the vagaries of metaphor and the stability of community, the evanescence and persistence of performance, the anonymity of spectatorship and consumer transaction over against the intimacy of real and vicarious community. In small places, through the accumulation of small exchanges, technique binds strangers to one another, transforming contingent arrangements, over time, into relational infrastructure.

2
Corporeal Chronotopes: Making Place and Keeping Time in Ballet

> *The beauty that we see in the vernacular landscape is the image of our common humanity: hard work, stubborn hope, and mutual forbearance striving to be love. I believe that a landscape which makes these qualities manifest is one that can be called beautiful.*
> (John Brinckerhoff Jackson, *Discovering The Vernacular Landscape*, 1984)

> *There are perhaps many histories of ballet that remain to be written, histories that have nothing to do with Sadler's Wells, the Ballet Rambert or the influence of Diaghilev.*
> (Lesley-Anne Sayers, 'Madame Smudge, Some Fossils, and Other Missing Links: Unearthing the Ballet Class', 1997)

> Work as hard as you can. Make it beautiful. *How can you argue with rules as pure and as simple as that?*
> (Sigrid Nunez, *A Feather on the Breath of God*, 1995; emphasis in original)

The protocols of dance technique do more than just construct readable, reproducible bodies in diverse urban communities. They also rewrite bodies' and communities' relationships to space and time, and to the intersections of both. Over time, technique creates 'vernacular land-scapes' within urban environments. Technique recreates neighborhoods as sites of productive, diverse allegiances. It transforms the sparseness of studios into home places and bodies into maps. It organizes relationships across culture and class to form affective environments, geographies of the heart. The vernacular landscapes constructed through dance technique are literal and psychic spaces for the daily, routine time and talk that bind practitioners to one another. As J. B. Jackson argues, such landscapes are always local, regardless of the ideals incarnated there; they are stabilized by idiosyncratic ways of seeing the world (1984: 149) as well as by the community's 'distinct way[s] of defining and handling time and space' (1984: 150).

Jackson suggests that such landscapes are not 'political,' in the sense of engaging those superstructures that 'official' discourses erect to maintain 'the public' (1984: 150). Yet, just as 'most buildings can be understood in terms of power or authority – as efforts to assume, extend, resist, or accommodate it' (Wells in Hayden, 1997: 30), just as there is, in effect, a politics of use, vernacular landscapes made through dance are deeply and thoroughly political. These landscapes are shot through with contestatory notions of appropriate gender performances and gendered resistances, culture- and class-inflected expectations of the relation-ship between art and life, and issues of discipline and authority. All of these are negotiated using technique as a road map for navigating family pressures, as a stage for the self, or as a refuge from routine. Technique constructs intimate, familiar places for a politics of self- and community-building that Hawes characterizes in spatial terms; this politics 'disclos[es] ways of escaping from and relocating to different subject positions at the same time [it] redraw[s] ideological boundaries' (1998: 273). The landscape I describe here is created in and through the daily labors of ballet, a seemingly unlikely confluence of 'elite' and 'vernacular.' But, as Lesley-Anne Sayers observes, 'the "meaning", imagery and associations of ballet in amateur practice' have their own unique and peculiar histories that do not relate 'simply and straightfor-wardly to ballet as an art form in the theatre' (1997: 135).

For the past 15 years, I have observed ballet classes, rehearsals and performances at Le Studio, a highly regarded dance training facility in Pasadena, California; for more than a decade, I took class there as well. Le Studio was founded in 1979 by Philip ('Phip') Fuller; he and his twin brother Charles ('Chip') serve as co-directors, both of the educa-tional program and of DanceCorps, its resident company.[1] The Fullers had distinguished dance careers, most notably with Pittsburgh Ballet and Ballet West, where they worked under the direction of Willam Christensen, one of the great pioneers of American ballet.[2] Gilma Bustillo, Le Studio's new Associate Director, joined the school after a distinguished career that included soloist roles at the American Ballet Theatre, and a principal position with John Neumeier's Hamburg Ballet. Le Studio's ballet faculty includes Julia Ortega, former soloist with Ballet de Camaguey, who defected to the United States in 1993 and joined the school in 1998.

Le Studio is located in a working-class neighborhood, part of northeast Pasadena. The residents are primarily African American, though a marked increase in the Latino population over the past ten years has made it more diverse. The area, filled with brightly painted bungalows whose windows are heavily barred, is anchored by a large post office and mail distribution center. Le Studio is part of a small, low-slung, modern office complex housing a software analysis and management firm and facilities for the State of California. A free-standing building, it covers 6000 square feet and includes three studios, communal dressing and rest rooms for both men and women, a lobby and an office. Ballet classes, all with pianists as musical accompaniment, generally take place in Studios 1 and 2; visitors are able to watch class by peering through the doors and windows of these individual classrooms or, more rarely, by sitting inside. Chip, Phip, Gilma and Julia all teach in the ballet program, which ranges from Basic I through adult classes. They are often joined by guest instructors and choreographers, who work specifically with the advanced students. Enrollment in the Student Division of the Ballet program averages about 300 annually. In addition to ballet, Le Studio offers classes in Jazz, Modern, Hip Hop, Flamenco, Tap and Pilates mat work.

In one sense, Le Studio is typical of the surprising juxtapositions that characterize global cities generally: a unique concentration of talent, difference, and sophisticated cultural infrastructure. Yet it is not a typical urban neighborhood ballet school. Its affiliated semi-professional company, the high quality of its staff, the intense involvement of its advanced students, its demographic diversity, and the large number of men and boys who train there set it apart. This exceptional dedication to, and high level of, ballet training makes it an especially rich site for observing how ballet, arguably the *uber* technique, organizes space and time in/as community for its students.

In the words of Xavier Roncalli, who was worked with the Fullers for 15 years and is now Director of Operations at Le Studio Fitness, the satellite Pilates facility a few blocks away, the culture of both Le Studio and its associated company is 'a great social experiment.' He means this jokingly, a reference to the range of idiosyncratic characters that have moved through over the years but, in fact, the demographics of the ballet program are far more diverse than conventional thinking might suggest. Students are working to upper middle class; scholarships and work-study opportunities are available for promising candidates who cannot afford class fees. When I began to take class at Le Studio in the late 1980s, students, including company members, were primarily Anglo. This has changed dramatically; both populations are now representative of the surrounding area, including Latino, Armenian, Mexican, African American, Chinese, Vietnamese and multiracial dancers at all levels. As noted below, the male corps of advanced students had been, until recently, older and more diverse than the largely adolescent female corps. This too has changed over the years as the average age of the men has decreased and diversity among the young women has increased. The men in the company have always been accorded special status among regional amateur and semi-professional companies; the large number of 'guys' is highly unusual for a local studio. Yet this 'critical mass' was limited to the advanced classes. This is no longer true. The population of boys in the lower divisions has increased over the years; at a recent spring school demonstration, there were seven boys in Basic I, the lowest level of formal training, and at least one, though usually two or three, at each subsequent level.

Le Studio, and its associated company DanceCorps, present two major performances during the year. The first is *The Story of the Nutcracker*, a shortened version of the full-length ballet.[3] The second is a Spring Gala and class demonstration involving students at all levels, from Threshold (pre-dance) to Advanced; this event typically includes performances by the company. The mission of DanceCorps is 'to increase the creative life of the community by encouraging future generations to participate in and/or experience learning through the medium of dance' (lestudiodance.com/dancecorps.html). To that end, the company does a number of school demonstrations and performances during the academic year. While the entire school participates in major performances, company members take on the major roles and responsibilities for these additional demonstrations. Many dancers move from the school to professional careers in companies, including those of international acclaim. Yet, as its mission statement indicates, Le Studio does not emphasize commitment to ballet as a career. Still, the school's record of excellence led the Willam Christensen Trust Foundation to award DanceCorps the exclusive rights to perform his ballets in the Los Angeles area, a high honor and an unusual one for a semi-professional company.

Like any school, part of the 'social experiment' at Le Studio involves setting young bodies into the sturdy matrices of technique, discipline and skill. Like any vernacular landscape, the experiment also includes ways of redeploying these matrices to serve individual and collective ends very different from any official mission. There are students who just pass through this social experiment and those who stay; Xavier observes that, after so many years, he 'seems to be a lifer.' As they all keep time and take their places at the school, and in ballet, they share the 'hard work, stubborn hope, and mutual forbearance' Jackson describes. They imagine and remake intersections of space and time, bodies and the vocabularies and protocols in place to receive them. Along the way, they may also remake self, community, and the generative spatial and temporal possibilities of inhabiting ballet in the global city.

Part I of this chapter examines and theorizes relationships between technique, place, and time in ballet. The technique is deeply gendered, both in theory and in practice. The two remaining parts of this chapter focus on what their places and times at Le Studio mean for the young women and girls (Part II), and men and boys (Part III) who train there.

PART I BODIES IN TIME AND SPACE: CORPOREAL CHRONOTOPES IN BALLET

Keeping time

Dance and time are intimately linked; in ballet this linkage is especially acute. Indeed, ballet technique makes the embodied experience of time communicable. Steps are counted into the future, beginning 'on the ones' and proceeding through 'five and six and seven and eight.' Classes are organized ritualistically, beginning with barre exercises and ending with *reverence*, a bow or curtsey to the instructor and, if present, the musical accompanist as well. Advancement in technique is tied to age and measured by 'levels'; in the Cecchetti method these are actual 'grades' assessed by exams, though, in many studios, these levels are enforced less programmatically.[4] The limits of the body are temporal markers: too young for pointe shoes, too old for quick jumps or fast footwork. The technique is so demanding that, even at the moment of mastery, it begets a backward glance: 'When I was younger, lighter, speedier ....' Time, for those committed to ballet, always seems to be running out: training should not start 'too late,' generally not after age 10, and careers are over 'too soon,' in a dancer's thirties or, with injuries, even younger.

The repertoire itself is replete with backward glances: to royalty and dancing peasants, and to Romanticism, both historically and more broadly construed. In recent popular culture representations, ballet's relationship to time gets radically truncated. Only two ends of the temporal continuum are dramatized: ballet as initiatory vehicle into woman- or manhood (*Billy Elliot*), or as monastic refuge for those who have aged in the clutches of technique without the apparent resources to do anything else (*The Turning Point*).

Yet, whatever its general contours, details of the relationship between ballet and time are always local, always contextual. Time in/and technique is very different for an American Ballet Theatre principal dancer, for whom the demands of career and chronology are intimately and fairly publicly connected, than it is for the thousands of girls and boys, men and women, taking classes in recreation centers, neighborhood studios, and performing with semi-professional companies. In these contexts, participants may dance toward dream time ('When I am a ballerina...'); they may dance away age ('I took classes as a little girl...'), paradoxically embracing it at the same time ('Can't jump like I used to'); or they may deploy ballet to carve out time, and identity, for themselves away from the scrutiny of parents, from the demands of a job, from more conventional ambitions. In each of these locations, in myriad idiosyncratic ways, communities of dancers deploy and circulate rituals and stories, enabled by technique, that help them both examine and navigate relations between the body, agency, and time.

Taking, and making, place

Likewise, dance technique re-visions and recreates space by literally 'placing' it in dialogue with the body. Space, writes de Certeau, is a practiced place (1984: 117). Space is multivocal, characterized by perpetual possibilities for transformation. Place is univocal, stable, proper. I would argue that the construction and reproduction of place from space can be explored in performative terms.

Judith Butler characterizes the performative as the 'power of discourse to reproduce effects through reiteration' (1993: 20); these 'effects' constitute 'identity,' which repetition then stabilizes. While Butler is reluctant to claim rhetorical power for the theatrical ('Performative', 1990: 278), Elin Diamond argues for a more generative relation between performance *per se* and performativity, one especially conducive to reading places of performance and performance's placement. She writes: 'When performativity materializes as performance in that risky and dangerous negotiation between a doing (a reiteration of norms) and a thing done (discursive conventions that frame our interpretations), between someone's body and the conventions of embodiment, we have access to cultural meanings and critique' 1996: 5). As useful as Diamond's reformulation of the relationship between performance and performativity is, I would put this another way. Individual performances, I suggest, do make performativity material but such negotiations are not always risky; they may be, or seem, perfectly banal. The performative production of place is one example.

Cultural meanings of place in dance, including the most sublime and the most familiar, are generated by iterations of bodily doings, organized by technique's protocols, its 'discursive conventions' for reading corporeality and geography (the thing done). That such doings may be banal rather than obviously subversive makes them no less meaningful or constitutive. Home places can be performatively stabilized through performances of the banal. Yet, as Diamond suggests, there are also other possibilities: chances to undo place, to enact erasures of technique's protocols containing 'doings,' to challenge sedimented meanings in performance and disrupt the readability offered by the thing done, or even feign that readability to insinuate something else. In such operations, space might be performatively generated from place. As geographer Yi-Fu Tuan observes, 'Place is security, space is freedom; we are attached to one and long for the other' (1977: 3). This performative negotiation of space into place, and the reverse, is implicit in Michel de Certeau's geopoetic essay 'Walking in the City.' He writes:

> The long poem of walking manipulates spatial organizations, no matter how panoptic they may be: it is neither foreign to them (it can take place only within them) nor in conformity with them (it does not receive its identity from them). It creates shadows and ambiguities within them. It inserts its multitudinous references and citations into them (social mores, cultural norms, personal factors). Within them it is itself the effect of successive encounters and occasions that constantly alter it and make it the other's blazon.... (1984: 101)

In her 'Introduction' to *The Geography of Identity*, Patricia Yaeger asks, 'Are there "rules" if not laws driving the narration of space?' (1996: 4). Are there regulations that dictate protocols for the performative transformation of space into place, or the reverse? In ballet, as in walking, the answer is both yes and no.

There are in fact some 'laws' by which ballet transforms space into place and these are explicit. The technique performatively constructs place by, first, 'placing' mapped bodies into it. These bodies are themselves viewed as spaces to be organized by technical protocols, then performatively stabilized. Even in 'pre-ballet' classes reserved for children aged 3–6, students are inserted into the Euclidean imaginary of the technique: shoulders and hips form the four corners of a 'square.' I remember quite clearly as a child, flushed with newfound knowledge of polygons, asserting firmly that the corporeal map so etched would, in fact, be a *rectangle*, only to be abruptly and unequivocally dismissed with the declaration that the body must be 'square' because 'squares are better than rectangles.' This was no idle geometric prejudice; it was a strategic one. The square, with its four equal sides, captures the rigorous metaphoric symmetry prized by the technique. Squares are always perfectly balanced; ballet dancers are to achieve the same. Bodies stay 'square' by ensuring that the hips and shoulders remain aligned and in the same plane during the execution of a movement. The illusion of two dimensionality is the ideal, as in 'turn out,' which rotates the upper thighs and, in turn, knees and toes away from the body so that, when feet are positioned with heels together, they form a line at a perfect 180°. An egregious violation of this symmetrical two dimensionality proves the rule: during *plié*, where turned out knees bend over the toes, the torso must be perfectly straight up and down. A rear-end sticking out beyond the heels, or a chest leaning forward beyond the toes, is an object of reprimand and ridicule, parodically labeled 'squaté,'; the ugliness of the phoneme and its scatological connotations clearly marks the violence of symmetry involved in improper corporeal 'placement.'

The body in ballet is 'centered'; movements originate from the torso, which is 'pulled up,' always poised between yielding to and resisting gravity, occupying a metaphoric 'middle place' of stability and balance. The center is also a reference point, a corporeal prime meridian running down the torso lengthwise: turning 'outside' (*en dehors*) is to move away from this imaginary axis, while an 'inside' turn (*en dedans*) initiates an action toward it. Arms and feet are mapped onto five positions; all movements begin in, go through, and end in a position, theoretically if not literally. Further, unlike de Certeau's walkers in the city, practitioners whose improvisatory negotiations of the landscape rewrite it in the most protean and anonymous terms, ballet dancers are given seven basic movements that govern interactions with space: one can bend at the knees toward the ground, rise on the balls of the feet or toes away from the ground, jump or stretch into space, turn around in it, or glide or dart through it.

Rules for corporeal placement in ballet are echoed by maps organizing studio space. Walls and corners are numbered; there are eight positions or spatial orientations possible in a room. Dancers in class may, as directed, face corners, generating a diagonal alignment, or walls, appearing straight on. The proscenium stage is the architectural paradigm for studio space; 'front' is where the audience will be, though this spot is occupied by a mirror in most studios, transforming 'front' from a dialogic location between dancer and audience into a 'screen wherein dancers enact [and display] their competence' for themselves (Sadono, 1999: 164). Thus, the mirrored front of the studio is both a model of the performance stage, the goal, and a metaphoric and pedagogical ' "stage" on which the dancer enacts her acquisition of the ballet ideal,' and through which she must pass, as she engages ' "the sociocultural construct of an ideal image" reflected back to her' in the process of training (Sadono, 1999: 165).

While the paradigm of the proscenium figures prominently in the logic governing ballet's negotiation of space and place, it is important to note that the studio is simultaneously configured, both geographically and chronologically, as onstage *and* backstage. Dance barres typically line three of four studio walls, exempting only the front or audience position. One-half to two-thirds of a ballet class takes place at the barre. Here, drills of basic movements reinforce steps and skills to be reassembled later into choreography. Warming up and stretching happen here. The barre is 'backstage,' where isolated mechanics are perfected. Generally, a time break separates barre work/backstage from center work/onstage, which emphasizes the combinations of steps intrinsic to choreography in performance before the gaze of the mirror/audience.

The ballet studio is a panoptic place; even the barre as 'backstage' is a place of surveillance by instructors, and self-surveillance by dancers looking in the mirror at (some would say 'for') themselves. There are also 'back backstages,' escapes from ballet's official panoptic regime. These are waiting and dressing rooms, though they, like the barre, may be tactically appropriated for self-display beyond the rubric of official pedagogy and for audiences other than the mirror or instructor.

Of his improvisatory walkers, de Certeau observes: 'The walking of passersby offers a series of turns (*tours*) and detours that can be compared to "turns of phrase" or "stylistic figures." There is a rhetoric of walking. The art of "turning" phrases finds an equivalent in an art of composing a path (*tourner un parcours*)' (1984: 100). While 'ordinary language' certainly has its place in the ballet studio, there are official 'turns of phrase,' strategic rhetorics of movement, that narrate the technique's geography. These turns are, literally, *tours*. Ballet technique writes its strategic ambitions onto bodies and/in space in French. If, as de Certeau implies, space can be configured in linguistic terms, then the use of French enacts a shift, not only in geography but also in time. The here and now of contemporary American dance practice is also there and then; ballet's historicity, its 'foreignness,' even its royal alliances are spoken into space and onto bodies in an 'other tongue.' This is not to imply that French in ballet transforms a studio in Detroit or Pasadena into another place in any literal or simple way. It is to suggest, however, that language explicitly signals the studio as a place of artifice, of a formality of relations, a place where bodies in the present step figuratively into the shoes of dancers from the past, even if the specifics of these other dancers are not known or invoked.

And yet, for all of the vocabulary regulating body placement, despite the labeling of walls and corners, and the use of a language that signals a ritual frame and historical continuity with dancers of the past as much as it serves utilitarian communication in the present, the answer to Yaeger's question of laws narrating space is also 'no.' Whatever ballet's strategic ambitions for bodies in/and space, individuals inevitably seize opportunities to narrate studio space, and performatively construct place, differently, in turns of physicality and of phrase that insinuate themselves into and redeploy ballet's rituals of proprietary power.

In 'Place in Fiction,' Eudora Welty writes: 'Location is the crossroads of circumstance, the proving ground of "What happened? Who's here? Who's coming?" – and that is the heart's field' (1978: 118). Likewise, Alan Gussow observes: 'A place is a piece of the whole environment that has been claimed by human feelings' (1971: 27). The ballet studio can be read in similar terms. In this view, labors of love and loss – affective, relational labors – performatively instantiate places for the heart as surely as *battements tendus* at the barre create places in and for the properly aligned body. Overlaid onto, and hidden within, the rigidity of the technique is a complex geography of fantasies: dreams of physical and social virtuosity shot through with longing for an autonomous agency that both masters and exceeds technique; flirtations, real or imagined, with partners or potential partners, also real or imagined; wistful reflections forward to 'grown up' pointe shoes or *pas de deux* class, or backward to younger days, quicker moves, more flexible muscles in other studios or in the same one. There are also geographies of talk: speculations on how technique will be mastered in performance in another place or, looking backward, how it was or was not, finally, brought off here or there. Remember *that* night, in *that* theatre, or in *that* class? 'Stories,' de Certeau writes, 'thus carry out a labor that constantly transforms places into spaces or spaces into places' (1984: 118), security into freedom and the reverse.

Ballet is particularly amenable to geographies of longing. Even the strategic narratives of the classical repertoire reinforce this: longing for a lost beloved (*Swan Lake*), for a favorite toy made whole (*The Nutcracker*), for freedom from filial, or royal, or banal duty (*La Fille Mal Gardée*). Yet there is a more personal and simultaneously more grandiose terrain that is longing's object in the ballet studio, one visited and lovingly recalled by every dancer who has forged a home place through the performative power of technique.

Over my years of observing generations of young ballerinas, and dancing myself, I have often been asked, always by non-dancers, what participants 'get' for their enormous investments of time, energy and, often, money. Ballet demands hours of classes and rehearsals per week at the same time adolescence demands its own performances in very different venues. As I will discuss below, for young women who wear pointe shoes, ballet also means pain, a high risk of injury and, in the long run, often ugly feet. There is little question of parental pressure to submit to ballet's demands; if anything, this pressure generally goes in the opposite direction. As I will discuss below, ballet is emphatically *not* the career path most working- and middle-class parents want for their girls. What these young women 'get' from ballet is varied; among other things, this is a matter of where they get 'to.' In addition to new places they may visit, new bodies they may inhabit as technique remakes them, new communities into which they are inserted, they gain access to what Jill Dolan calls 'the intersecting worlds of performance and pleasure,' (2001a:15) to places of both enormous self-consciousness and complete self-erasure. This is what Gaston Bachelard, in *The Poetics of Space*, characterizes as 'intimate immensity,' an 'attitude [or an affective geography] that is so special' that those in its grip are transported 'outside the immediate world to a world that bears the mark of infinity' (1964: 183).

Here, at this intersection of corporeal and affective places, Bachelard observes, 'One feels that there is *something else* to be experienced besides what is offered for objective expression. What should be expressed is hidden grandeur, depth. And so far from indulging in prolixity of expression, or losing oneself in the detail of light and shade, one feels that one is in the presence of an "essential impression" seeking expression . . .' (1964: 186; emphasis in original). The place of intimate immensity is both destination and reward for dancers. Here, as Erica De La O (see below), now a professional ballerina, whom I first met at Le Studio over a decade ago explains, a dancer is swept up in 'the rush of it all coming together: the work and the beauty and the love you have for it [the technique]. It all comes together.' This is a place of sublime culmination, paradoxically a utopia – a 'no place' – performatively constituted from the daily labors and daily longings demanded of and inspired by technique.

For Bachelard, 'intimate immensity' is a solitary place (1964: 183); in ballet, it is not. It cannot be. Even illusions of autonomous agency in ballet require others: musicians; a *corps* of 'supporting players,' background against whom one's genius shines more brilliantly; partners as props to be 'lifters' or 'liftees'; audiences, even if rendered synecdochal as clapping hands. Once, as I was pulling away from a regional ballet festival in Modesto, California, my car full of young men from Le Studio who had performed there, I was mobbed and brought to a halt by a swarm of girls, all of them probably between 12 and 14. They were clamoring for one of the young men to get out of the car, which he did with the palpable air of indulgence and *noblesse oblige* one would expect from a rock star assuaging rowdy but adoring groupies. Two queues quickly formed, rotating girls between positions as photographer/documentarian and documentary object. These girls wanted to be lifted, held by this young man in a pose known as 'the fish' (or, when carelessly executed, 'the death fish'). One by one they jumped into his arms. He grasped each by her extended leg and torso and, with her other leg bent in *passé*, one arm demurely crossed at the chest and the other extended backward in arabesque, he theatrically dipped so that her upper torso seemed to skim the asphalt. Yet the smiles and the clicks of the cameras, the giggling rotation of these queues – as precise as the lifts themselves – seemed to dissolve that parking lot. We were at Lincoln Center, or the Dorothy Chandler Pavilion, or the Modesto, or Fresno, or Paris equivalent. It was not the young man who was the object of longing here; the girls did not want pictures of him *per se*, or even of each of them with him. He was, rather, the prop necessary to sustain the Prima Ballerina, the Soloist, at her moment of apotheosis in her glorious, imaginary geography.

Ballet technique offers utopian spaces and home places where the limits of small-town parking lots, or suburban strip mall storefront studios, or recreation centers in global cities expand, performatively becoming grand stages for bodies and wishes, suffused with applause. But it may also instantiate densely oppressive dystopias that implode into black holes of eating disorders, Darwinian cut-throat competition, and plain authoritarian meanness, all glossed over and held in place by invocations of technique's (or 'art's') demands. Injuries are also places: radical compressions of space/time written into the body when days or years of training condense and congeal into a strain or a break, a twist or a pull. Even in nurturing studios, and certainly in non-nurturing ones, it is important to know one's place in terms of gender or hierarchy.

In ballet, spatial practices, like chronological ones, are locally enacted. However they may be legitimated by appeals to transhistorical precedent or ahistorical purity, space and time in the studio are uniquely inflected by the bodies and stories that put them in play. Moreover, like bodies and stories that inhabit them, spatial and temporal operations generally, and in ballet in particular, are intimately linked in a kind of experiential relativity. As Henri Lefebvre observes: 'Space is nothing but the inscription of time in the world, spaces are the realizations, inscriptions in the simultaneity of the external world of a series of times . . . (1996: Kofman and Lebas, 16).

Corporeal chronotopes

In 'Forms of Time and of the Chronotope in the Novel,' Mikhail Bakhtin theorizes time as a social location. The chronotope, he argues, names 'the intrinsic connectedness of temporal and spatial relationships' (1981:84). Here, '[t]ime, as it were, thickens, takes on flesh, becomes artistically visible; likewise, space becomes charged and responsive to the movements of plot, time, and history' (84). Chronotopes are plural; they describe organizing and support functions of representation and meaning, much as technique serves as the infrastructure shaping corporeality and narrating space in dance. Further, chronotopes, like technique, are prospective, delimiting 'a field of possibilities' and specifying 'the possibility of events' (Morson, 1994: 106).

Bakhtin developed the chronotope to examine historical poetics in literature, though he also offered possibilities of more expansive readings; these are especially useful for understanding dancing communities in the global city. He writes: 'Out of the actual chronotopes of our world (which serve as the source of representation) emerge the reflected and *created* chronotopes represented in the work (in the text)' (Bakhtin, 1981: 253). Thus, the chronotope functions as a dialogic intersection of the world and the representational grammars and protocols that organize and reproduce it, like technique. The metaphors Bakhtin uses to characterize the relationship between the 'actual world as sources of representation and the world represented' (1981: 253) are generative. He writes:

However forcefully the real and the represented world resist fusion, however immutable the presence of that categorical boundary line between them, they are nevertheless indissolubly tied up with each

> other and find themselves in continual mutual interaction; uninter-
> rupted exchange goes on between them, similar to the uninterrupted
> exchange of matter between living organisms and the environment
> that surrounds them... Of course this process of exchange is itself
> chronotopic... We might even speak of a special *creative* chronotope
> inside which this exchange between work and life occurs....
>
> (Bakhtin, 1981: 254)

Bakhtin's organic imagery here recalls his other physiological metaphor used to characterize the chronotope: where 'time takes on flesh' (1981: 84). As Clark notes, 'Time assuming flesh is something more than a trope here, for those who enflesh the categories are people' (1984: 280). More specifically, they are bodies.

I suggest that chronotopes are corporeal as well as textual. They are enacted by material bodies that invigorate formal, representational grammars and protocols like dance technique and, in turn, manipu-late these grammars and protocols for their own ends. Here the ballet studio, in all its temporal and geographical specificity, becomes the membrane facilitating this chronotopic process of exchange between particular bodies and the technologies of representation, between art and life.

Two additional aspects of the chronotope make it especially useful for characterizing the busy intersections of bodies and technique, space and time in the ballet studio. In *Dialogism: Bakhtin and His World*, Michael Holquist notes an apparent 'antinomy' in the concept: 'Certain chro-notopes are treated by Bakhtin as if they were transhistorical structures that are not unique to particular points in time. There is a tension, if not a downright contradiction, between these examples and the claims Bakhtin makes elsewhere for the chronotope's ability to be in dialogue with specific, extra-literary historical contexts' (Holquist, 1990: 112). Holquist suggests resolving such a contradiction by thinking of the chro-notope 'bifocally': 'invoking [the chronotope] in any particular case, one must be careful to discriminate between its use as a lens for close-up work and its ability to serve as an optic for seeing at a distance' (1990: 113).

Holquist makes a valid point but corporeal chronotopes in ballet suggest thinking differently about 'bifocality'; perhaps less discrimina-tion rather than more is called for. As I mentioned above, ballet tech-nique – its *lingua franca*, its grammar of bodies and/in space, and its repertoire – places its participants in a 'both-and' relationship to here and now, there and then. In ballet, participants literally follow in the transhistorical footsteps of other dancers, though their pointe shoes may be made from high-tech composite materials rather than layers of fabric and glue. Both the convergences and the discrepancies between here and there, now and then matter, and the corporeal chronotope of ballet holds them in tension, not as antinomies but as partners.

As a formal category, the chronotope links the organizational mechanics of telling (*szuzhet*) to the messiness of the told (*fabula*); it is the place 'where the knots of narrative,' those intersections of events and plots, 'are tied and untied' (Holquist, 1990: 113; Bakhtin, 1981: 250). Corporeal chronotopes in the ballet studio inhabit and enact these same dynamics as they performatively constitute space and time. Here, ballet's technical protocols of reading and writing the body generate 'formal and rational design' in class and 'enslavement to narrative' onstage, particularly in the action ballet 'war horses' that so often figure in the repertories of semi-professional companies (Foster, 1998: 258, 262). Yet corporeal chronotopes also organize the 'just-so stories' of self-creation in community, stories that redeploy ballet's technical rhetorics to other, infinitely more idiosyncratic ends (Clark and Holquist: 1984: 68). For example, they emerge in those banal and essential moments of reminiscence where events retrospectively become narrative: 'Remember when?'

Two corporeal chronotopes are particularly useful for describing how ballet organizes bodies, time, space and stories; I call them 'roam' and 'home.' 'Roam' is a performative version of Bakhtin's chronotope of the road; it reflects the process of moving, as well as the location of movement, that characterizes the corporeality of time and space in ballet. 'Roam' refers to: '. . . both a point of new departures and a place for events to find their denouement. Time, as it were, fuses together with space and flows into it (forming the road); this is the source of the rich metaphorical expansion on the image of the road as a course: 'the course of a life,' 'to set out on a new course, [or a course of training.]' (Bakhtin, 1981: 244). Moreover, the road may be 'one that passes through *familiar territory*, and not through some exotic *alien* world' (Bakhtin, 1981: 245; emphasis in original). In ballet, the chronotope of 'roam' often unites both the familiar and the exotic.

To roam is to move away from 'home,' narrowly construed as the domestic sanctuary of/for the family, perhaps toward a new home in technique. Both popular representations of ballet and the chestnuts of the classical repertory bear this out. Pursuing ballet means leaving home, whether such leave-taking involves moving across town or to another part of the country (*Flashdance*; *Billy Elliot*). Typically in these accounts, the terminus of roaming, like Bakhtin's road, is a point where 'social distances collapse' (1981: 243), if only, as in these examples, through fantasies of pure merit which admit the hardworking seeker to the exotic and hallowed inner sanctum of technique. Likewise, the repertory is filled with images of roaming: Clara and her Nutcracker Prince journey to the 'Kingdom of Sweets,' where familiar confections become dancing marvels; Albrecht goes slumming incognito in a peasant village and Giselle travels back from the dead; James follows the Sylphide into the literal and affective woods.

Roaming is, in fact, logistically central to a professional ballet career; frequently this involves moving, or aspiring to move, to New York, where not one but two highly esteemed companies beckon like the Promised Land. It may mean, in the words of one former Le Studio dancer, a man in his late thirties who left to pursue a peripatetic career, 'hitting the road with [or between] the farm clubs,' the relative minor leagues and small markets of companies across the country. Even with a position in a comparatively secure major company, roaming means touring, not only in the Unites States, but across the world as well. Most dramatically, chronotopes of roaming in ballet link geopolitics to bodies, as in the famous Cold War defections of Rudolph Nureyev, Natalia Makarova and Mikhail Baryshnikov, or in less publicized contemporary dancers' flights from Cuba.[5]

Yet, just as all movements in ballet begin, go through, and end in a position, all roaming revolves around, starts from, or ends at home. Bakhtin frames this as 'the space of parlors and salons': '. . . this is the place where encounters occur. . . . In salons and parlors the webs of intrigue are spun, denouements occur and finally – this is where *dialogues* happen' (1981: 246; emphasis in original). Ballet offers all the comforts and discomforts of home. In the classical repertoire this is literal: ballets are set at home (*The Nutcracker*), involve literal or fantastic flights, or attempted ones, from home (*La Fille Mal Gardée*), or clearly establish whom one can or cannot make a home with (*La Sylphide*; see especially Banes and Carroll, in Garafola, 1997). These are often uncanny places, 'unhomey' (*unheimlich*) homes, haunted by giant rats, vengeful Willis, dancing swans, or lifelike wind-up dolls who beguile suitors. Freud's exemplary narrative of the uncanny at home is E. T. A. Hoffmann's 'The Sandman,' the basis for the ballet *Coppélia*.

Ballet companies and ballet studios are also home places; they may be every bit as haunted as the domestic spaces in the repertory, or they may be refuges where dancers go to escape from personal or familial demons. Relations between ballet and home are multi-faceted; studios may have all the formality of the parlor or all the lived-in familiarity of the den. Erica, introduced above, tells of being at home with, and in, technique: 'My body is at home with it because it's so familiar and yet it can always surprise me.' Studios and companies can reproduce familial relations, for good or ill. Sasha Anawalt's description of an exodus of dancers from the Joffrey Ballet is one example; dancers left 'because [Robert] Joffrey expected his dancers to act as if they were his children, subordinate themselves to his will, and not question him on any matter, artistic or bureaucratic' (1996: 204). Anawalt concludes: 'Joffrey quickly realized that the only people who would fully accommodate him were kids' (204). In *Dancing On My Grave*, Gelsey Kirkland describes working for George Balanchine in familial terms: 'After the death of my father, I turned with grim determination to the face that seemed to hold my future...I accepted the sentiment that was awakened in my benefactor [Balanchine], recognizing that surrogate paternity was preferable to the kind of inclination he displayed toward other dancers he favored...Mr. B. treated me like a wayward daughter...' (1986: 53). She eventually characterizes her relationship with Balanchine as a kind of patriarchal hell.

Obviously not all metaphoric familial configurations in ballet are this dramatically dysfunctional. Still, the interpersonal stakes in a ballet-home may be high. When two members of the Le Studio staff divorced, rumor had it that two of the men in the company, longstanding 'regulars' in their late thirties whose own childhoods were framed as 'difficult,' 'couldn't take it.' The fracture of their ballet family, according to this communal reading, was too shattering to manage and they left the studio, moving on to other things.

Ballet may literally be a family affair: Kirkland describes her (dystopian) relationship with her sister in Balanchine's company (1986: 58), but other examples are happier. Twin brothers Charles and Philip Fuller danced happily for the same companies; many students at Le Studio, both present and past, are siblings, often 'converting' recalcitrant brothers or sisters by bringing them to class or to performances. Xavier met his wife in adult ballet classes at Le Studio; his son and daughter are now taking classes there.

Bachelard notes that 'all really inhabited spaces bear the essence of the notion of home' (1964: 5). The ballet studio may not be a home place as a site of resistance, as bell hooks describes. Yet it is 'really inhabited,' offering both a social and an affective location that binds its constituents together in competition and in solidarity. Indeed, particularly for young women, as I argue below, the ballet studio does become a site for 'sharing feminist thinking and feminist vision' in ways both similar to, and very different from, hooks' view of the homeplace (in Madison, *Woman* 454).

PART II   HOME GIRLS: GENDERED SELF-FASHIONING AND SOLIDARITY AT LE STUDIO

This section examines the chronotopes, and the social, aesthetic and physical pressures and pleasures, young women encounter as they make places for themselves in ballet and at Le Studio. It begins with three accounts illustrating variants of ballet as home in Los Angeles. The technical protocols of classical ballet are deeply and heteronormatively gendered. These protocols are engaged, resisted and redeployed by Le Studio's female students to manage relationships to their male colleagues and parents, to pain and beauty in/of performance, and especially to one another.

Where the heart is: three stories of ballet in Los Angeles

Tensions between the chronotopes of roam and home in ballet are inflected differently depending on issues of gender and class. Such tensions clearly emerge in the case of Misty Copeland, a home town (Los Angeles) girl whose ballet career and 'family' battles became a *Los Angeles Times Magazine* cover story in December 1999. Popular culture has often demonstrated this tension between 'home' and 'roam' in ballet, generally in deceptively simplistic class-inflected terms, using the career trajectories of home girls and, with the release of the film *Billy Elliot*, home boys as well. In these representations, both the demonstrably, almost excessively meritorious, individual hero and the ossified, hallowed halls of 'official' ballet are redeemed. Sometimes the home boy/girl is self-taught, uniquely talented, sustained by only a dream against all odds. Sometimes s/he labors through some down-at-the-heels provincial training, usually at the hands of a kindly but odd instructor/surrogate parent who, often as not, had to cast her own dreams aside. Local oafs and doubting higher-class snobs must be managed. In the end, these 'authentic,' 'natural' dancers, plucky kids from the wrong side of the tracks, blow the cobwebs out of the fusty bastions of pinky-up high culture with sheer grit, street virtuosity, and their self-evident 'realness.'

The broad contours of Misty Copeland's story seem to fit these parameters so well that, at one point, 'Hollywood noticed the press coverage [of her situation] and swarmed' (Adato, 1999: 20). In 'Solo in the City,'

Allison Adato recounts Misty's 'discovery' at 13, an advanced age to begin ballet training. According to Adato, local ballet instructor Cindy Bradley was teaching a free class at the Boys and Girls Club in San Pedro, south of Los Angeles. She spotted a girl 'watching quietly from high in the bleachers [and] encouraged Misty to take a spot at the barre, wearing socks and borrowed gym clothes. That session prompted Bradley to send notes home [to Misty's mother], offering Misty a full scholarship' (Adato, 1999: 16).

Misty, '[t]his girl with the mocha skin, the product of African, Italian, and German heritage' (Adato, 1999: 16), was one of six children born to single mother Sylvia DelaCerna. In Adato's account, DelaCerna 'didn't like [Bradley] from the beginning... her elitist attitude, her snobbiness' (16). Misty's by all accounts prodigious natural talent set the stage for a battle between mother and mentor, each professing – sincerely it seems, in this account – to have the girl's best interests at heart.

After training with Bradley, transportation problems forced Misty's mother to end her ballet classes; to prevent this Misty moved from her mother's room in a residential hotel, shared with three other siblings, into the Bradley family home. Almost immediately, prestigious companies offered her places in their highly competitive summer programs. These were unheard of opportunities for a student with so little, and such recent, ballet experience. Misty left public school to study independently, allowing for full-time ballet training.

At some point, DelaCerna informed her daughter that she could no longer continue training and living with the Bradleys. Lawyers were called on both sides. Eventually, an emancipation request from Misty was withdrawn and she returned to live with her mother. She went back to high school, training only in the afternoons. Tensions over dancing continued, as Adato recounts:

> DelaCerna sounded at first as if she and Misty agreed... In minutes, however, she was urging her daughter to appear in 'The Chocolate Nutcracker' again. Misty protested. 'But they'll pay you,' said her mother. Misty took the part and earned $500.[6] To coordinate bookings, DelaCerna enlisted a family friend to act as her manager. ('All wrong,' said Gerald Arpino, artistic director of the Joffrey Ballet. 'She'll become a little gypsy. She'll learn the things that are commercially eye-catching and never know the art of performing as a great ballerina.')
>
> (Adato, 1999: 6)

Eventually, Misty was accepted into the highly competitive American Ballet Theatre (ABT) summer program in New York. ABT was her dream company; one of her heroines danced there. After exemplary work during that summer, ABT artistic director Kevin McKenzie asked her to join the studio company immediately. She demurred and asked her mother, who was back in Los Angeles unable to see her daughter perform because she couldn't afford to leave work. Her mother said, at first, that she would like to keep Misty home for another year but then observed, according to Adato, 'I've listened to what they are saying about Misty and I'm going to let her make up her own mind' (1999: 53). Despite the fact that ABT addressed her mother's concerns (they would arrange accommodations and schooling for her and pay her), Misty elected to return home to Los Angeles to finish her senior year in high school and stay with her family. 'I hope they still want me next year,' she said (Adato, 1999: 53).

Readers' responses to this story were sharply divided, each fixing on one fragment of this complex tale: 'Let's hope Misty's mother gets her own act together and allows Misty a starring future' (Sachs, 2000). Another reader wrote: 'Once many years ago, as a promising dancer myself, I was given an opportunity to train with the best. But between my own teenage insecurities and my parents' wish that I "have a good education to fall back on," I stayed put, finished my education and got a sensible, professional job. My dancing muse, disappointed in me, left and never returned...' (Grover, 2000). A patron who contributed to Misty's financial support countered: 'Misty's contentment with her life and her family situation is best summed up by her comment on the cover of the *Los Angeles Times Magazine* [on which she appeared] she autographed for her mom: "To my number one fan, my loving mother"' (Cantine, 2000).

The ultimate resolution of this story is less important, for my purposes, than the way its recounting and the responses it generated gloss the multiple chronotopes of ballet.[7] Taken together, the story and readers' reactions are a virtual allegory describing ballet's perils of home and roam for young women, especially those from working-class back-grounds. First, time is of the essence: 'the muse' might be unhappy and leave; the student might not be training hard enough; the teenager might miss the pleasures of adolescence. The nature of home itself is contested. If home is where the heart is, this shifts for Misty. Is it with her family? With her dream company? What place must she leave? Where will she return? Is ballet a path to a paying job, or a road to 'the art of performing as a great ballerina,' where 'commercially eye-catching' is a devil term? Where does promise, or regret, reside? Note that, if ballet is to be a professional home in any significant sense, Misty, like ballet dancers before and since, has to leave Los Angeles, which has no major resident professional company of its own (see Anawalt's account of the attempt to bring the Joffrey Ballet to the city, 1996: 325–36).

Misty Copeland may be a technical prodigy; accounts of her career may reveal the antinomies between roam and home in an especially dramatic fashion. Yet in other more general ways, her story is like that of many young dancers in Los Angeles. As Sayers observes, '[t]he aristocratic values associated with ballet [those implicit in Arpino's championing "the art of performing as a great ballerina" versus the merely "commercially eye-catching"] make a poignant contrast with the social context of its amateur situation and practice' (1997: 135). These amateur contexts are the birth and home places of many in the next generation of dancers. In these settings, as well as in the imaginary hallowed institutional halls of technique, dreams and bodies are made, managed and narrated. Though there are few *dei ex machinae*-like invitations to join ABT on the spot, and little *sturm and drang* caused by dueling lawyers, the stories from these amateur contexts attest to the generative possibilities of fashioning a home out of time, place and ballet technique. Erica De La O's story is one of these.

Erica is a beautiful woman in her twenties, the only daughter of Mexican immigrant parents, whom I had observed at Le Studio for over a decade. She has a perfectly oval face, delicate features, and the long and deceptively fragile neck and limbs of a ballerina. She is now a member of the Louisville Ballet, which she joined after successful seasons with the Columbia City (South Carolina) Ballet. She was accepted into this, her first professional company, after auditioning on a lark, going only to accompany a friend, figuring it was a way to take a new, free class. She accepted the soloist contract with Louisville after a successful appearance at the 2003 New York International Ballet Competition; she was selected to perform a solo at the Competition's closing performance at Lincoln Center (Shattuck, 2003; Anderson, 2003). She worked at Le Studio Fitness, the pilates studio (see Chapter 1) before leaving for the Company. At the end of her first year in South Carolina, in June, 2001, she returned to Los Angeles and danced the title role in Act II of *Giselle* in Le Studio's Spring Gala. She was so stunningly lyrical, so beautifully expressive, and her technique so flawless that those of us who had known her for almost a decade had to choke back tears. Little Erica! Remember?

I remember Erica as a coltish 'Tech II' girl at Le Studio, one of a small, select cohort of 11–13 year-olds who would be invited, on Fridays, to leave their regular class and venture into the studio next door, crossing the pedagogical and social threshold to enter the *pas de deux*, or partnering, class reserved for Company members, relative (and sometimes literal) 'grown-ups.' 'The guys [in class] loved us,' she recalled. 'We weighed, like, five pounds,' and so were ostensibly easier to lift. Her brother Pablo was in class as a member of the company. 'I was "Pablita" to everybody, just Pablo's little sister,' she recalled.

Erica is bright, focused, happy and self-reflexive. Her personal narrative reflects this. I recorded her story outside a local coffee house shortly before she left for South Carolina for the first time (25 August 2000):

Erica: I was very, very ambitious about my dancing when I was a little girl. I would put on shows for my parents. I'd move the furniture around and I'd dance and sing; I learned the words to Spanish songs, and my parents, of course, would be the audience. We lived in a little apartment building when I was growing up and my mom didn't want me playing in the street so she found out about this lady: Miss Baldwin, Martha Baldwin. She used to teach ballet behind her house in her garage .... So my mom took me there. We did the same *barre* every day; she lined us up from tall to small. There was a record player. She was an old, sweet, adorable lady, very Catholic. My mom fell in love with her and she fell in love with me. She guided our whole family into an American way of life. So I started dance there, in the back of her garage, on a cement floor. She ended up being my godmother; we were very, very close.

She opened people's eyes to art. I mean, who would want to teach ballet in that area? She also gave guitar lessons. She did this for $2.00. She came from a higher class in Mexico and she wanted to give. She'd put on these huge shows at the church. She'd buy costumes, the whole bit. It was cheesy in some ways but, for a kid, it was glorious.

Frank Martinez[8] started ballet with Miss Baldwin. She saw talent in him and she brought him to Phip and Chip. He loved her dearly; she was so lovely – just benevolent – and amazing. And he went to go visit her and take her class out of gratitude. My mom saw him and she said, 'Pablo, you should go see him. He's so good! You should take ballet too!' And Pablo was, like, 'That's for boys who are gay. No way!' Well, my mom always had us together, so she dragged Pablo there and he saw Frank Martinez doing tours and seven pirouettes, and he said, 'I want to try it.' So my brother got in there and Miss Baldwin fell absolutely in love with my brother: those long legs, so tall. So Frank saw us for a while and told Pablo about the studio [Le Studio] and he decided to train there. And my mom put me there too. I think Miss Baldwin was a bit hurt. It was a big commute for my mom: 40 minutes each way and Pablo was on scholarship so he had to be there every day. But it worked out.

My mother was dedicated to us; she wanted to expose us to everything she wasn't exposed to. Art was a big thing for her and, for me as a kid, ballet was a very different ambience. It's weird. Pablo got all these scholarships and then made the decision to quit. I didn't join the company at 13 like the other girls. My mom knew I fell in love with ballet, but she thought it would take too much time. I didn't fight for it then; I don't know why. Maybe I was scared. But once my mom saw Pablo's time there, she thought, 'Hmmm.' And then it was, well, I just had to go on.

. . .

I see two strands to my life. I want an academic career and a ballet career. After my car accident, I got very focused on both. I have Pilates training, and ballet will always be a priority, but I'd like to be a chiropractor. I really want to work with women: everyday women with no dance experience who want to stretch or feel flexible, and dancers too because I can communicate with them. It may be hard if my dancing continues, but this is the same discipline as dance and I'm going to be smart about it.

My mom reminds me of how far I've come. And they remind me – they don't want to spoil me, so they tell me about responsibility, about how hard this was for them and about how much this has cost. But they are totally, completely supportive. My mom is *there* and she is there in *the right way*. It's harder for my father because he doesn't relate to it as much. But he provides for me. He works so hard. And Pablo [now happily married and, at the time of this narrative, expecting his first child]! He thinks it's cool. It's so funny. He still does double *tours* [a jump from a standing position with two rotations in the air] in front of my face! He can still do splits. We still mess around – like ballet mess around – *tombé pas de bouré* [a basic, four-step sequence central to a young dancer's first combinations]. He reminds me of Phip and Chip. He's that kind of good man.

In *Choreographing History*, Foster argues that '[t]he organization of descriptive narrative can trace out the patterns and shapes that moving bodies make' (1995: 16). Though she is referring specifically to tasks of historical description and criticism as each attempts to represent the dancing body, her point also registers more broadly. That is, the organization of the personal narratives dancers offer to account for their lives in ballet likewise trace the contours and locations of movements linking technique and everyday life. Kristin Langellier reminds us that '[p]ersonal narrative performance is situated not just within locally occa-sioned talk – a conversation, public speech, ritual – but also within the forces of discourse that shape language, identity, and experience' (1999: 127). For dancers, these forces include the protocols of technique that make the body communicable and the corporeal chronotopes that insert these protocols, and these narrators, into space and time. Thus, in Erica's story no less than Adato's account of Misty Copeland, 'roam' and 'home' become nodal points, both anchoring and advancing the construction of identity in/and story and reshaping, claiming, perhaps re-enchanting or domesticating ballet in the process.

Adato's story of Misty and Erica's personal narrative share similar motifs; as noted above, some of these have been elevated by popular culture to a central place in the social imaginary of ballet. The first is the 'unlikely context' as ballet's first home place in the lives of these dancers, the place from which they must roam to grow and succeed. Misty is 'discovered' in a recreation center gym; she takes class in socks and borrowed gym clothes. Erica's first ballet home is Miss Baldwin's garage. Elements that seem purely descriptive in her account may also be read as an insider discourse communicating precisely how humble this form-ative home place was: cement floors, not wooden, sprung, or composite ones; a record player, not an accompanist; lining up tall to small and the same *barre* every day, odd rigors compared to students choosing whatever spot at the *barre* they want and varying exercises to alternately build and rest specific areas of young bodies. The logistics might have been 'cheesy in some ways,' but, in Erica's affective geography, they are transformed into something 'glorious.' Her mentor was 'adorable, bene-volent'; indeed, she became a *de facto* member of the family. For Erica, Miss Baldwin's benevolence was the tie that bound ballet to her family home; from her perspective, this was a happy, generative connection in contrast to the fraught scenario of Misty's mother versus her mentor.

Unlikely contexts as initiatory vehicles between home and roam in ballet are not unusual, even outside the simplistic characteriza-tions of popular culture. Consider the story of another home girl.

A *Los Angeles Times* article describes her career, her trajectory through roam and home, in fairytale terms: 'Once upon a time there was a little girl from Lynwood [a working-class suburb of Los Angeles] who wanted to become a ballerina. The little girl trained at a ballet school in Compton, then she trained in New York. She trained and trained until she became so good that a major company hired her. And she lived happily ever after – until she returned to Los Angeles' (Stewart, 2001: B1). In this account, African American ballerina Robyn Gardenhire roamed to New York as a member of ABT, among other prestigious companies. She left behind a supportive family who, like Erica's, encouraged their child's passion despite the financial and logistical sacrifices required of them to sustain it. A problem emerged when Gardenhire returned 'home' to Los Angeles: 'What she saw here would sour any dancer's storybook ending' (Stewart, 2001: B1). Los Angeles had no viable, professional ballet company, so Gardenhire founded the nonprofit City Ballet of Los Angeles. The company school was, and still is as of this writing, housed in the Salvation Army Red Shield Youth and Community Center in the Pico-Union section of the city, an economically depressed, densely populated area that is home to large numbers of Latinos, and Central and South American immigrants, whose children became Gardenhire's students. Her goal, she says, is 'to create a company that looks like Los Angeles': 'Ballet has always been a primarily Caucasian art form,' she said. 'That's only because our kids are not pushed toward that. If you don't see yourself in it, you don't emulate it. Once you see yourself, you can say, 'I can do that!' (Stewart, 2001: B14).[9] Underlying this story of unlikely contexts is a cultural solidarity at work to remake ballet in its own image. This solidarity also circulates through Erica's narrative. Miss Baldwin, an immigrant from 'a higher class in Mexico,' is a vehicle through whom other immigrant families can facilitate access to the arts, to 'a different ambience,' for their children. One of these 'children,' ballet dancer Frank Martinez, in turn passed this gift on to another generation. Martinez was instrumental in securing formal training for both Erica and her brother Pablo and more: he was a mirror of their ballet ambitions before they knew they had them. He was a cultural emissary for the form; indeed it was after seeing Martinez that Pablo observed, as Gardenhire predicts above: 'I can do that!'

Yet the question remains: why is ballet the object of these networks of cultural solidarity at all, particularly given Gardenhire's accurate characterization of it as 'primarily a Caucasian art form?' Why subject children to the hardships of 'roam' and the sacrifices from 'home' to generate new homes in the technique? Surely organized sports would keep them off the street as well; surely any 'cultural capital' ballet might muster, particularly in Los Angeles, must pale beside the infinitely greater visibility of both entertainment industry capital and, closer to home, maternal sacrifice. The answers to these questions are complex and they vary with the gender of the dancer; as I will discuss below, there is no parental unanimity on the kind of dividends paid by an investment in a daughter's ballet lessons. One simple answer is that once some children like Erica, Misty Copeland and Robyn Gardenhire 'see themselves' in ballet, they 'fall in love' and, in Erica's words, 'just have to go on.' Of course the dynamics supporting this love are also both complex and gendered; so is what these young women actually fall in love with, as the second half of Erica's narrative suggests:

I love the discipline – the focus – it affects you in life. It's a wonderful tool. I could not be distracted in school, especially when I started doing [ballet] competitions and performances. There's only one way to get through it. Plus there's the self-expression. It's great to have your mind have control over your body, to shape it, mold it. It's endless and it brings together so much.

For me, class is addicting and performance is the gift. When you're working, you have to deal with things internally. You push through so much. It's kind of like working with a psychologist, but on your own. You need to focus. You need to start thinking technique-wise, or stop thinking technique-wise. And you need to leave aside teenage stuff.

In school and in college I don't expose the ballet issue so much. It's not how I present myself. So many of my friends are in ballet – we complain about it, we don't flaunt it! There's a huge difference between my ballet and my school friends. Now my school friends are very aware but the ballet girls are *there*. We don't need to explain it: injuries, aches, we're in tears because our feet are dying and we've got two more hours of rehearsal. We give each other support. We share stories: 'I know. I've had that happen.' There's a huge connection. It's more intimate, especially when you're growing up.

It's actually so much easier managing teenage stuff with ballet. I don't know where I'd be without it. My girls, we were so close. Whether it was talking about sex or about school, about money, college, 'what are we going to do?' – We were on the same wavelength. Different backgrounds, different personalities, and we each pretty much found ourselves. It's something the ambience of the school

[Le Studio] brought together. We were allowed to be very friendly – very competitive – but we were dear to each other. It was a huge connection. HUGE. And if you'd feel that you can't talk to your parents about it, or you can't talk to your high school friends about it – because high school is such an intense place – they were there. It builds confidence. You meet so many other people and realize that we're all coming together but every single one of us is from a different city, different life, different culture and they'd bring their families and friends, and you'd keep meeting and eventually there's a huge vote of confidence in you. You're able to socialize with and respect different cultures and it's awesome and fun.[10]

In 'Dance Narratives and Fantasies of Achievement,' Angela McRobbie examines the popular culture motifs organizing dance, and particularly ballet, in the social imaginary. Her observations resonate with elements of Erica's narrative and help to address the specifics of students', particularly girls', affective investments in the technique as a home place and why, at least initially, families across class and cultural lines enable it. McRobbie writes: 'Art is something which can change the course of a life. It can give a young girl a legitimate reason to reject the normative expectations otherwise made of her and can provide her with a means of escape.... [D]ance is presented ... as something worth giving up a lot for, something that will pay off later in terms of great personal fulfillment.... [I]t acts as a participative myth, a fantasy of achievement and a way of taking one's destiny into one's own hands' (1997: 230). For McRobbie, the combination of romance and the importance of work explains dance's attraction as a tool of self-fashioning for young girls. It is important, though, to note how relatively little attention Erica gives to the 'romance' element in her narrative and how much is devoted to work. Performance is the 'gift' after the disciplined yet addictive labor of taking class. Sorority is forged, not through communal dreams of pink satin shoes and lacy costumes, but in those moments where 'we're in tears because our feet are dying and we've got two more hours of rehearsal.'

Young women at Le Studio and its affiliated company work hard. 'Tech II' students, those at the penultimate rank in the school, take an hour and a half of class four days a week. Advanced students, including most company members, take an hour and a half of class five days a week. During performance season, roughly October through June, there may be four to six hours of rehearsal on Saturdays, often with Sunday and weeknight rehearsals as well.

Parental investments, cultural capital

Like Erica, most of the young women in Le Studio's company whom I've spoken with and observed over the past decade were brought to ballet by mothers whose initial ambitions were modest. Ballet was an outlet for energetic girls who loved to perform, awkward girls who might develop better coordination, girls 'dragged' to class to accompany a sibling already enrolled, or girls who were, at least in their mothers' opinions, in want of something safe and structured to do other than playing in the street. Many of these girls began classes around age 6, with some starting earlier or later. Early parental support was enthusiastic, in marked contrast to the ambivalence many parents, mainly mothers but also some fathers, have expressed about their daughters' (and, much more rarely, sons') involvement in the company.

There are generally three categories of parents at Le Studio; I think the school is fairly typical in this regard. The first category supports and finances their children's ambitions and attends performances (and sometimes classes) with minimal additional overt input or, from the children's point of view, minimal interference. Erica's mother (and Robyn Gardenhire's) is one of these. She is, as Erica says, '*there*, and she's there in the *right way*.' She generally lets Erica make her own decisions in light of a clear understanding of her responsibilities and her family's sacrifices on her behalf. Erica's father 'doesn't relate' to ballet, but clearly loves and supports her. Note that implicit in Erica's characterization of her mother is the possibility of 'being there,' as a ballet parent, in the *wrong* way.

The remaining two categories of parents express the strongest ambivalences about their daughters' intense commitments to ballet, coming from completely opposite positions. These are the parents who may not be 'there' in quite 'the right' ways. In Adato's account, and in some of the responses it generated, Misty Copeland's mother seems to oscillate between these two. One of these groups consists of 'stage parents,' again generally mothers, who have professional ballet ambitions for their girls whether their daughters share these ambitions, or share them with the same intensity, or not. None of the communally designated 'notorious' parents in this category at Le Studio would speak to me directly about these ambitions, or about their mechanics for fulfilling them, but they did figure prominently in dressing-room gossip. Rumors about whose parents called Phip and Chip to threaten, cajole or bribe a daughter's way into a part circulated through small groups of girls, particularly prior to *Nutcracker* auditions in the fall. These rumors were always impossible to verify. By all accounts, Chip and Phip never give in to, or reveal, sources of parental interference. In the dancers' parlance, such pressure 'rolls off their backs,' and they are respected and admired for this. I did once overhear one young woman, a daughter of particularly egregious stage parents, moan to her friends, 'This is so – God – SO embarrassing' during one rumor-swapping session. She was met with compassion, perhaps because she was a sunny and genuine young woman, perhaps because her parents were so notoriously 'weird' by community standards that they engendered some sympathy for their daughter among her peers. As one member of the group whispered to another after the embarrassed girl left, referring to the rumored parental interference: 'Can you IMAGINE?! I'd, like, DIE.'

The second category of parent was more forthcoming about concerns with ballet as a daughter's home place. Generally, though not always, these were upper-middle-class mothers and fathers whose investments in ballet's cultural and disciplinary capital moved in the opposite direction from those of the stage parents. Ballet was acceptable as a means to an end, not as an end in itself. While stage parents were concerned about too few performance opportunities, this second category of parent was concerned with 'too much.' 'Mrs A'[11] insisted: 'It's too much. It's too much, just too much. It's so isolating, you know? How will she meet people? This is her senior year. She should be out meeting people, having fun. But all this time here – Of course I'm really proud, you know. But it's, it's too much.' 'Ms B' voiced similar concerns: 'I'm so glad she's out [of the company – the young woman was still taking class]. This way she got to, you know, be a cheerleader and be in Student Council and be in clubs. When — [a former ballet mistress at the school] wanted them to miss their proms – their *proms*! – that was, you know, too much.'

This ambivalence, and the recurring invocation of excess, of 'too much,' are multi-valenced and merit attention. Note that even Erica's supportive mother demurred when she wanted to join the company at age 13, fearing that it was 'too much.' Certainly at least some of these parents' concerns are financial: too much money. In addition to class fees, pointe shoes for young women are expensive: $45–$65 a pair. A dancer may go through three pair a month, more during a performance run. Add to this more nominal and irregular, but still significant costs for dance clothes and doctor bills in case of injury and a family's financial investment in a young dancer is considerable.

All of the parents I interviewed spoke of their delight in their early recognition of their daughters' talents; they described the girls as artistic and creative children and so they were more than willing to make sacrifices for ballet lessons as an 'outlet' for these, particularly, as one upper-middle-class mother observed, 'given the shoddy nature of the arts in public school.' Lessons were seen as the supplement to a well-rounded education. That the outlet was beautiful and at least nominally affiliated with 'classing up' was all the more reason to begin lessons. 'It's so beautiful and expressive,' said one mother as she watched her daughter take class. 'I mean we love to *go* to the ballet but, after all this time, you have to think about what kind of payoff. . . ' She never finished her sentence.

Ballet's 'payoff' is complex for this group of parents, and its cultural capital is leveraged against another valuable commodity: 'too much' time. Bourdieu is right to suggest that distinctions between elite forms like ballet and popular ones reflect and reproduce existing power relations. This is particularly true for the 'dominated fraction of the dominant class,' at pains to distinguish itself from the 'popular class' through practices of social distinction (1984: 316–17). Bourdieu, however, does not sufficiently distinguish between the cultural capital generated by *consuming* distinction ('We love to *go* to the ballet.') and that generated, or diffused, by actually using one's labor to *produce* it (being a ballerina).[12] As another mother observed: 'Ballet's given her great posture and great discipline but you have to wonder if her wanting to do this so much isn't sort of diminishing returns. I try to tell her that SC [University of Southern California] doesn't give ballet scholarships. At least, I don't think they do. And even if they do, that means you pretty much have a job. You have to dance even more. Which is great, I guess,' she hastened to add, 'if that's what she really wants. But what about time for college life and time for a career?' Those practices of distinction, 'great posture, great discipline,' must be assessed in light of yet more markers of status: a prestigious university, a career (versus 'a job'). Perhaps, this mother intimates, there won't be time for both. While self-expression, and parental recognition of its importance, is certainly one marker of class privilege here ('– if that's what she really wants'), its relative priority is only ambivalently asserted. There are no scholarships for ballet, probably, and if there were, they must be purchased, perhaps, at the price of both college life and career by the time consuming labor of 'dancing even more.'

This parental ambivalence 'at home' can also be examined in light of the class affiliations of ballet which, like those of the Khmer Apsaras, are more complicated than popular culture representations and invocations of royal lineage would suggest. Louis XIV is credited with codifying ballet in the West but, after the French Revolution, the form's status emerged as much more equivocal (see Lee, 2002, for a general history of ballet). In Europe, the Romantic period ushered in pointe work and, along with it, a new 'feminization' of the art. While this 'uniquely female utterance' (Garafola, 1997: 4) in/as ballet elevated some dancers to international stardom, it had other unintended consequences as well. As Garafola observes: '[w]here pointe ushered women into a realm of perpetual maidenhood, an idealized version of the separate female sphere, [ballerinas' assumption of] male dress with its leg revealing tights, announced their sexual availability. Indeed, in the Romantic period, when family-centered training largely gave way to institutional-ized training in academies, the connection between ballet dancers and certain forms of prostitution deepened' (1997: 6).

Janina Pudelek writes of the 'seraglio' aspects of the ballet school attached to the Wielki Theatre in nineteenth-century Warsaw. As it did not charge tuition, the school's student population 'came, almost without exception, from poor families' (Pudelek, 1997: 143). Women students and dancers used their positions with the ballet to secure 'rich protectors' or 'marriage prospects in the world of craftsmen, merchants, and civil servants' (Pudelek, 1997: 147). Pudelek notes: 'Although insistent moral and financial pressure may have driven some ballet dancers to stray from the path of proper conduct, the great majority regarded the question of "additional income" as a necessary evil, and if there was no compulsion, knew quite well how to reject such advances' (147). Likewise, in the United States ballet had a dubious reputation, linked to music hall 'leg shows.' As Judith Lynne Hanna observes, ' "ballet girl" had a pejorative connotation in the US until the mid-twentieth century and, in some places, it still does' (1988: 124). It is only relatively recently that ballet recuperated its high culture allegiances. Interestingly, this class slippage and subsequent rise is directly linked to the absence, then resurgence of pre-eminent, well marketed male dancers.

There is, then, a great deal of parental support for engaging in practices which would mark a young dancer as feminine and bourgeois, so long as these do not locate the girl outside the circle of 'meeting people' and 'the prom,' that is, 'too much' outside the mainstream. There are other investments 'at risk' here beyond the financial. Even the most ardent stage mother in the company, still notorious in the community after almost a decade, affirmed these concerns as she sized up her daughter and available companies for a match. 'Mrs C' fretted: 'Well S-F [San Francisco Ballet] is auditioning, but they're all gay. The men are all gay. And Joffrey – I don't know if they'd take her, she's so short. Maybe if they needed a Clara [for *Nutcracker*]. And, you know, that's the trouble too: who's she going to meet? Who's she going to date? Those Joffrey guys, I checked it out: they're all bisexual.'

Here, as with 'Mrs A,' this is not an issue of how a young woman will meet people; clearly she will do that. This is an issue of how she will meet the *right* people. For this set of parents, the diversity Erica so enjoys about her ballet home may be a useful learning experience en route to a middle-class life, but it may not be a desirable destination like 'the prom' or 'SC' where 'meeting people' will 'pay off' in terms of networking, career and class appropriate 'dates.'

Other parents express concerns in terms of simple utility, investments or economics, using formulations like the reference to 'diminishing returns' above. 'What will she do with all this work?' asks one mother. Adato observes that '[I]n most circles, going to college is a measure of achievement; among hopeful young dancers, it is what happens when you don't make the cut' (1999: 20). Parents in this category, though, have other ideas. The low-paying, highly competitive, peripatetic life of a dancer is not what they want for their girls. This 'work' is not producing the kind of, or desire for, a future they envision as an acceptable index of achievement. Thus, the concern becomes 'too much' wasted time on 'all this work' of dubious ultimate use value: too little return on the time invested. Significantly, this issue emerges precisely at the temporal pivot when a young girl's disciplined diversion, acceptable for instilling core values, turns to become, instead, a late adolescent's serious commitment, putting parents' view of the appropriate deployment of those values in jeopardy.

It is important to recall here, as noted above, that despite the sizeable number of alumns who have gone on to careers in ballet, Le Studio does not demand or expect that students and company members have professional ambitions. Both Charles and Philip Fuller have university degrees in fields other than dance; both continually emphasize the importance of good grades and a well-rounded, happy life as keys to positive experiences with ballet. Indeed, this is one of the reasons parents are drawn to the school. As one father told me: 'I've got to hand it to these guys. They're right there with asking about grades, always talking about being smart, good in school, how important that is. Not like [another local studio] with this attitude [nose in the air, caricaturing effete snobbery], you know, "You must sahhhcrifice because only ahhht means anything." These guys are real with these kids and I respect that.' Erica likewise confirms that 'the guys are great role models. They're good businessmen, good teachers, good dancers. They show how to put it together.' Thus, the exigency undergirding these parents' anxieties is not some external pressure on their girls. As another father observed, 'There's no guru-Svengali crap here.' Rather, there is instead the disconcerting possibility that parents' capital investment in a class-appropriate diversion, one that shores up discipline and focus, might be diverted by a daughter willfully opting out of a class-appropriate future, choosing 'ahhht' and 'sahhhcrifice' over what's 'real.'

> '*Mrs D*': I think about how serious she is and about all this time: years, really, all these years. What will she do with it?
> *JH*: Well, don't you feel the discipline she learns here and the other less tangible benefits will transfer over to other things?
> '*Mrs D*': Transfer over? Well, yes, yes. They'll transfer over, but let's be real. They won't transfer over to a first-rate law school.

Consider another example of this class-inflected relationship between art and 'being real': an excerpt from Annie Dillard's *An American Childhood*. Dillard writes, 'My friend's father was an architect.... He had been a boy who liked to draw ... so he became an architect. Children who drew, I learned, became architects; I thought they became painters. My friend explained that it was not proper to become a painter; it couldn't be done. I resigned myself to architecture school and a long life of drawing buildings. It was a pity, for I disliked buildings' (1987: 80).

If distinction is acquired and reinforced by going to the ballet, it is also potentially, and dangerously, dissipated by, as one mother put it, 'slaving away' at producing it. Beyond instilling generic virtues of discipline and focus, the ability to do four pirouettes with a single preparation won't transfer to law school, or to the law firm's box seats, season tickets and eligible bachelors.

Hiding in the light

> ... *a notion of contradiction* must *be brought to bear in any attempt to understand the full complexity of women's relation to culture* ...
> (Tania Modleski, *Feminism Without Women*, 1991; emphasis in original)

The girls, of course, are familiar with these arguments. Some hear them nightly. The question becomes: in the face of very real physical and emotional complications involving their commitments to ballet, why continue? Why try to forge a home place from technique, as a potential career or as an intense personal investment of time and energy, when those in your other home may have other plans? One answer comes from 'Gina,' daughter of 'Ms B,' who left the company at the beginning of her senior year in high school: 'What I miss, what I really miss is that when I was doing it, no one could tell me, could *seriously* tell me, "You should be doing something else." It's not like hanging out, though we do hang out. No one can really complain that much about it. I mean, what you're doing hurts. It hurts and its hard and not everyone can do it, so no one is gonna seriously say, "That's stupid. You should do something else, something important."' Another dancer chimed in, 'I mean, after ballet, what's "important"? What're you gonna do instead, be a cheerleader?'

Ballet, then, becomes a way of securing a space of relative leisure and community against the prying of relatives and teachers for whom adolescents' free time must be weighed against 'what's important' and perhaps, what is both 'productive' (transferring to law school) and gender and class appropriate: being a cheerleader. Despite some parental concerns about 'too much,' girls are able to subvert conventional pressures to engage in 'productive' activities and 'hang out' with a diverse group of friends in ballet without appearing to 'just' hang out. Values of discipline, hard work, and pursuit of individual excellence congeal into what another dancer called 'my alibi,' an unimpeachable one, a nearly foolproof strategy for what Dick Hebdige calls 'hiding in the light.' Hebdige characterizes 'hiding in the light' as a process whereby young people deemed 'problematic' convert surveillance designed to contain and control them into an assertion of visibility, into 'the pleasure of being watched' (1988: 8). Though he writes about British punks, delinquents and, for contrast, the white leisure class as indexed in mass media, his observations have striking relevance to American girls in the ballet studio.

Ballet is scopophilic; not only is the product of technique consumed through looking at a finished performance but, as the prominence of mirrors in studios attests, the final display in performance is produced through a meticulous and relentless process of visual scrutiny that is panoptic in its ambitions. The instructor's gaze is internalized and redeployed as the subject body examines herself in the mirror, assessing her visible articulation of an abstract ideal. The result is an internalized feedback loop as the dancer continually enacts and adjusts to self-surveillance that will ultimately lead to display before an audience. But more than the utility and pleasures of visibility are redeployed as young dancers hide in the light of ballet technique. The disciplinary underpinnings of that visibility are also used by these young women as 'alibis' against the more conventional expectations of parents and, as Erica observes, against 'teenage stuff.'

As 'Gina' noted above, whatever the nature of parents' concerns, they cannot object to ballet as 'unimportant'; the discipline required, and clearly and visibly invoked by the technique, elevates it to a level of import greater than the comparatively trivial role of 'cheerleader' that pales in comparison to the focus demanded by ballet. Because of its very obvious physical and mental demands, ballet becomes a legitimate way to 'hang out' in solidarity with young women from a wide range of backgrounds (and men, about which more below). If ballet is to be construed as a waste of time by parents investing in specific trajectories of achievement for their daughters, the hard work involved in this form of hanging out means that parents, not the children, are on the defensive. As one Tech II girl observed when her mother raised objections about the time involved in her moving up to the company, 'I just said, "Okay, no problem. More time to hang at the mall." It was like she couldn't buy me pointe shoes fast enough.' This girl's knowing smile clearly indicated that she recognized the rhetorical winning hand and was willing to play it. Ballet versus hanging at the mall: it was a parental no-brainer.

Yet it would be a mistake to conclude that ballet's discipline is, for these girls, simply a way to leverage some measure of autonomy from parental expectations. Indeed, discipline is the first thing Erica mentions when asked to explain her love for ballet. Parents see the technique's emphasis on focused self-control as generic; indeed, in late adolescence it is too generic, not 'transferring over to law school.' For the young women themselves, however, this discipline is specific and instrumental, a way of consciously fashioning an embodied identity against those discourses conspiring to impose one on them. As 'Gina' put it: 'When you're [on pointe] thinking "How can I do this without, like snapping your ankle in half, and can you do it every night [of a performance]," you're not into who gets some designer purse. It's like "Who cares? How pathetic. Get a life."' The full weight of consumer culture is brought to bear on these adolescent girls; in this discourse, you are what you buy. Yet, within the home place of ballet technique, McRobbie observes, 'the physical body seems to be speaking in a register of its own choice' (1997: 230). Further, the choices involved are both incremental and longitudinal: do I choose this form of training to mold my body and, as Erica says, 'affect [me] in life'? How does this hour of practice build to quicker turns or more endurance? These are not the immediate and superficial choices of choosing identity via selecting one brand of product over another.

Registers of choice in ballet are ambivalent. I am not suggesting that the form provides young women with unmitigated free agency for infinite and multifarious self(ves)-fashioning. To paraphrase Judith Butler in *The Psychic Life of Power*, to thwart parental injunctions to produce docile, class appropriate bodies is not the same as dismantling or changing the power-laden terms of subject constitution (1997: 88). It is more accurate to suggest that, in choosing ballet as a place to hide in the light from other expectations – whether those of parents, of consumer culture, or of conventional 'teenage stuff' – these young women choose one disciplinary regime over another.

Ballet, the regime they choose, offers the opportunity for these young women to see themselves as authors of both their corporeal and their social beings, even as they are aware of how they are authored by the technique. As Erica observed in her narrative, 'It's great to have your mind have control over your body,' to author one's own physicality, when popular and parental discourses are eager to do it for you.[13] Further, as Erica notes, in ballet the embodied self is both something you 'find' through engaging different bodies and backgrounds and something you 'work on' through your own labor. Self-expression is, in Erica's view, 'a plus,' important but secondary; it is the effect of, not the purpose of, this self-fashioning through this 'wonderful tool,' this discipline and focus that 'affects you in life.' To access self-expression, the dancer must move beyond simple mastery and explicitly and reflexively redeploy, even transcend, the vocabulary that authored her: 'You need to start thinking technique-wise or stop thinking technique-wise.'

In *The Uses of Pleasure*, Foucault characterizes self-fashioning in aesthetic terms, using the formulation 'the arts of existence': 'What I mean by the phrase are those intentional and voluntary actions by which men [sic] not only set themselves rules of conduct, but also seek to transform themselves, to change themselves in their singular being, and to make their life into an *oeuvre* that carries certain aesthetic values and meets criteria' (1985: 10–11). For young women at Le Studio, ballet is simultaneously a performing art and an art of existence, a way to 'shape and mold' the body and the mind: 'It's kind of like working with a psychologist, but on your own.' Further, this process of self-authorship in ballet is not finite. Like arts of existence, it is processual: 'It's endless and it brings together so much.'

Yi-Fu Tuan observes that place functions as 'an analeptic – a creative solution' (1992: 39) that allows its constituents to temporarily and provisionally 'forget our separateness and the world's indifference' (44). Place grounds arts of existence within the global city, contextualizing them as surely as the stage organizes arts of performance. Among other things, the dailiness of ballet and the ballet studio as home, as the venue for endless refinements of body and mind, bring together a community of young women in competition and in solidarity, onstage and off. Indeed, competition is transmuted into solidarity as 'Marie,' whose twin sister was also in the company, explains:

> With the twin thing – it's like there's always competition and closeness anyway because you can't get away from each other and you want to be, you know, to have what's *yours* sometimes but have that closeness too. So I've always had that but I didn't get it [understand it] until ballet. Because an 'A' is an 'A' you know? But in ballet, everyone gets something and nobody gets everything – well the perfect star does but here it's more: 'She's got the turn out and she's got the extension and I can turn.' So you always try to be better, to get more than what you naturally have, but nobody has yours so you work together so everybody gets better. And everybody knows it so you don't have to say it.

'Marie' always 'had' a sense of the intricate couplings of competition and closeness; in her ballet home the technique and the solidarity it generates through daily practice allow her to 'get it,' both in this arena and in her family home with 'the twin thing.' This kind of awareness, encompassing an individual's capacity to author self, the sense that she is also authored, and the self-reflexivity to operate on her own construction generates an *esprit de corps* that, as Erica notes, exceeds even camaraderie with 'school friends' who 'are very aware.' Ballet technique produces ties that bind without having to be said, even if much about this camaraderie is, in fact, spoken, in/as stories, and performed as community rituals above and beyond whatever gets told and performed on stage.

It is interesting to briefly compare the camaraderie forged in the ballet studio to a more general portrait of American girls in this same age range, 13–17. As Joan Jacobs Brumberg argues in her review essay, 'When Girls Talk,' that portrait conveys 'a level of contemporary adolescent unhappiness and demoralization that seems overwhelming' (2000: B8). Little, apparently, has changed since Lyn Mikel Brown's and Carol Gilligan's landmark study, published in 1992 as *Meeting at the Crossroads*. Brown and Gilligan found that, among other positives, girls navigating adolescence became more autonomous and less egocentric:

> Yet we found that this developmental progress goes hand in hand with evidence of a loss of voice, a struggle to authorize or take seriously their own voices in conversation and respond to their feelings and thoughts – increased confusion, sometimes defensiveness, as well as evidence for the replacement of real with inauthentic or idealized relationships. If we consider responding to oneself, knowing one's feelings and thoughts, clarity and courage, openness and free-floating connections with others as signs of psychological health ... then these girls are in fact not developing .... (1992: 6)

Further, the girls in this classic study 'were finding themselves at a relational impasse; in response they were sometimes making, sometimes resisting a series of disconnections that seem at once adaptive and psychologically wounding: between psyche and body, voice and desire, thoughts and feelings, self and relationship '(Brown and Gilligan, 1992: 7). As Gilligan observed, girls are 'confident at 11, confused at 16' (in Brumberg, 2000: B7). Brumberg notes that feminist scholars' concerns with issues of voice and adolescent girls have recently been transmuted into a spate of texts allowing readers to 'listen in' (2000: B8) on young women's concerns which seem, if anything, even more fraught than those Brown and Gilligan described. While these works do address the dearth of girls' voices in popular culture and in social science, Brumberg suggests that the proliferation of highly emotive memoirs and personal narratives may be symptomatic of a larger cultural failure to bring about systemic change in girls' psychosocial development. She writes, these 'antidotes to silence': '... are obviously part and parcel of an entrepreneurial and confessional culture that can be lurid and tawdry. That culture ... explains the emotional tone and the highly individualistic politics of such books. ... Despite the claim to be insurgents, most young women trust naively in the ethic of rugged American individualism. An autonomous, individuated girlhood appears to be the answer to all their problems' (2000: B10). Against this backdrop of self-censorious disconnections and isolating individuality, what Erica calls 'teenage stuff' and the 'intensities' of high school, her testament to 'her girls' who were, and are '*there*,' her plans to become a chiropractor to share her knowledge of, and passion for, the body with 'everyday women,' and 'Marie's' observation that 'you work together so every-body gets better,' stand out in high relief. Of course, not every girl is as focused, positive and socially invested as Erica or 'Marie,' though they are more typical than not. That said, in over a decade of observing young women at Le Studio, I know of only one young woman seemingly done in by the challenges of adolescence. This girl's tribulations were met with repeated, collective interventions by her ballet friends. These interventions continued even after she left the company and, more dire to her peers, stopped coming to class. When these efforts eventually proved unsuccessful, the young women involved were still wistful and saddened by the situation, keeping tabs on their former colleague, some even after a decade. She was the exception that proves the rule of sorority and solidarity at Le Studio.

While it may be overreaching to suggest that ballet inoculated these young women, protecting them from the social and psychic ravages of sexism at the crossroads of adolescence and adulthood, it is clear to me that, in addition to serving as a kind of home place, Le Studio is also a refuge, a safe and generative harbor, an analeptic as Tuan suggests. Here, in addition to deep and authentic friendships, girls gain both voice and community in the process and from the fruits of their individual and collective labors. Further, as Erica suggests, in the time and talk woven through these labors, they receive 'a huge vote of confidence.'

Certainly some of the credit for this goes to Chip and Phip, and to Gilma, Julia and other ballet instructors who have, over the years, offered the girls a variety of models for living as independent, expressive women in and out of dance. But much of the credit goes to the girls themselves, and to their ability to read life through ballet and the reverse. As 'Patty,' the ebullient 14-year-old daughter of a working-class Vietnamese refugee family, observed: 'I don't feel like there's a wrong way here. Maybe because I'm in the middle. There are some older girls who dance with us and I see what they're doing – and the guys – and I can look back at the Tech II to see those girls and where I've come from. It's very safe, in a way – to try things, to work on things. The only really bad thing is to be lazy and not try.' 'Patty's' place 'in the middle' between those men and women who train at Le Studio and the Tech II girls where she 'came from' recalls Brown and Gilligan's 'crossroads.' Yet, for 'Patty' and her peers, this is not a place of disconnection but one of balance, of equipoise between individual excellence and group solidarity, body and mind, between, to paraphrase Soyini Madison's words, the pains and the pleasures of effective creation.

On her toes: pains and pleasures of creation in community

In her essay, 'Oedipus Rex at *Eve's Bayou*,' writing on a film that also features girls at a crossroads, Madison argues that the pleasures and pains of effective creation 'do something worthwhile on contested ground,' like adolescence, 'while, at the same time, they are emotionally worthwhile insofar as we are affected and feel something that is unforgettable' (2000: 315). As art's practices and products organize 'new landscapes for the imaginary' (314), she notes: '[y]ou see something created, you hear it [and embody it] and it stays with you for better or worse. It is effective for you and it affects you. It becomes a political and emotional encounter. It refuses silent appreciation because it is not simply appreciated; it is more complicated than that. It is something that needs to be worked over, questioned, implied, imposed, perhaps accused and sometimes praised. But the most important thing, the most poignant thing, is really about connections' (Madison, 2000: 315). When the adolescent girls at Le Studio deploy ballet as an art and an art of existence, these connections, pleasures and pains, and their corporeal and emotional half-lives, are never far from their talk. They don't 'flaunt' their training outside the studio, as Erica indicates, but they do joke about it, rhapsodize about it, and complain about it among themselves.

In ballet, particularly for women, the pains of effective creation are literal. This is true for creating selves through arts of existence and for creating performances onstage. Discomforts, whether minor or profound, are aspects of both arts that must be 'worked over' and worked out, as Madison suggests. Indeed one of the rules of conduct governing ballet as a home place and as a communal art of existence involves the public management of pain. A popular 'Fullerism,' a formulation used teasingly by Chip and Phip and routinely repeated as a signal of insider status by Le Studio students, is 'If it feels good, you're doing it wrong.' This Fullerism is tongue-in-cheek; it is also generally true. However well trained and warmed up the dancer, ballet can be, and often is, painful. A study released in 2000 concluded that professional ballet dancers have the same frequency and severity of injuries as athletes playing contact sports (Nagourney, 2000). Yet, as 'Susie,' 'Marie's' twin sister observes, 'When everybody's legs are hurting, nobody wants to hear whining about it in class. Just work it out. We're all hurting and we all get through it.' As one dancer put it, 'Pain is background noise. In rehearsal especially you don't want someone calling your attention to it. You just deal.' This is what Erica means when she says, 'When you're working, you have to deal with things internally. You push through so much.'

Yet there are limits to what is dealt with internally. No one wants to hear whining in class or in rehearsal but, afterward, pain moves from 'background noise' to front and center in these young women's talk. Della Pollock suggests that 'pain doesn't precede language, that it is always already interwoven with meanings and meaning systems.... It travels' (1999: 121). For young women at Le Studio, pain circulates as a kind of interpersonal currency. As Erica observes, swapping stories about pain and sharing rituals for dealing with it is 'intimate.' Participation in these stories and rituals generates a very specific kind of connection, a solidarity forged from both empathy and endurance. 'Gina' observed: 'Between matinee and later shows [of *Nutcracker*], we'd go to somebody's [another dancer's] house and cut our tape off [some dancers tape their toes when wearing pointe shoes] and pull the skin off the blisters – I know, it's gross – and we'd all stick our feet in the tub and pour a lot of alcohol in so they [the blisters] would dry up for, before the next show. It was – I guess it was really sort of fun, actually, in a sort of gross way. Like battle wounds, you know?' Watching the girls take a break from rehearsal, I overheard 'Mara' admonish her younger sister 'Lily,' who was not used to more intense pointe classes: 'No! Don't pull on it [a toe nail]. Don't pull it off. It'll turn black and it'll come off. It'll just come off. See these are [two of her toe nails are black.] I know it's sort of weird digging in there, but it'll fall off and the new one will just be there under it. Just don't touch it.' As I looked up from taking notes on this exchange between sisters, this connection of blood and blackened toe nails, another young dancer, rubbing her feet, caught my eye and smiled. 'I know,' she said, nodding toward the other two girls, to her own feet, and to a small, nearby pile of pointe shoes. 'It's like, so weird doing this in a way. Like dancing on blocks.[14] But we're all in it together so everybody gets it and it's beautiful in the end – hopefully.'

Elaine Scarry suggests that pain is an isolating experience; to have pain is to 'have certainty' but to hear about it is 'to have doubt' (1985: 4). Yet here 'everybody gets it' and 'we're all in it together.' This solidarity in pain is, even more specifically, a sorority, a connection forged through the synecdoche for all the contradictions young women negotiate in ballet: the pointe shoe.

Young women in dance keep on their toes, poised at the crossroads of various disciplinary formations and modes of surveillance, between individual achievement and corps community; on pointe they pivot between beauty and considerable discomfort. 'Gina,' who noted above that she 'really missed' having a kind of moral authority over her own time while in ballet class, was alternately adamant and wistful when I asked her, years later, what she did not miss. She was a junior in college then, and a dance major. She observed: 'I couldn't stay away! But I'm doing modern now. No more pointe shoes – ever, ever, ever! And I do not miss it at all. It's like my feet are out of jail! Except sometimes when I see a ballet, I do think, "Wow. How beautiful and long legs look on pointe!" Sometimes bare feet just feel clunky and sloppy and just too close to the ground. Does that make any sense?'

'Gina's' ambivalence is not idiosyncratic; it permeates every aspect of pointe work, including its history. Though, as Jennifer Fisher notes, no one knows exactly who invented the pointe shoe (1999: 64), women were at the forefront of its development as part of what seems, from a contemporary perspective, a kind of feminist project: to literally and metaphorically elevate women above men who held the pre-eminent positions on stage. Garafola characterizes the use of pointe shoes in the mid-nineteenth-century as a 'weapon in the ballerina's growing arsenal' (1997: 4); Chazin-Bennahum notes that dancers like Amalia Brugnoli were exploring the potential of pointe shoes in the early 1820s (1997: 124). But, according to Judith Lynne Hanna, it was Marie Taglioni in the 1830s who 'established a foothold for women [by] employing the toe dance as an essential element of ballet' (1988: 125), leading to the celebration of the technique as 'feminine aesthetic quintessence' and even 'a revulsion against male dancers' (125). Hanna continues: '[w]hile the tight fitting toe shoe...restricts natural movement and perpetu- ates the ethos of female frailty and dependence on male authority...it also permits the dancer a range of movements, positions, and height impossible in other footwear. In a sense, the toe shoe raised women above the herd and out of the house' (1988: 125). Only women work on pointe. When men do so, it is for the explicit purpose of appropriating the feminine position for parody as in the drag company Les Ballets Trockadero de Monte Carlo.[15]

As 'Gina' suggests, pointe shoes generate a beautiful, long line of the leg, and an escape from clunking on the ground, even as they feel like podiatric jail. They are also not without risk. While good teachers do not put students on pointe earlier than ages 11 or 12, and reputable retailers won't sell the shoes to the untrained, injuries happen even under the best supervision. Though many problems are transient, they can include inflammation of the tendon to the big toe and osteoarthritis (see Nagourney, 1999; 'Sunday', 2001). Falling off pointe shoes could lead to ankle sprains and even breaks. In the short run, blisters and loss of toenails are also consequences of work on pointe. Finally, pointe requires considerable strength. Dancers must be 'pulled up' to take as much weight as possible off the feet; legs and ankles must be capable of both enormous power and great flexibility to enable quick rises on toe.

Girls in Le Studio's company, even those as young as 14, are quite aware of the ambivalence of pointe work as well as the enormous amount of energy expended on, and the strength and discomfort required for, movements coded and read by audiences as 'fragile' and 'ethereal.' They are also fully aware that pointe shoes, and the ways they are read by audiences, are gendered and political:

> *'Joanna'*: I hate how the guys [in the company] whine, 'Oooo tours [*tours en laire*, a turn in the air from a jump off of both feet, landing in the same position]. 'If they had to do *fouettés* on point [a turn requiring the dancer to go from flat to pointe on one leg while the other moves from full extension front, to side, to *passé* at the knee. On seeing the girls execute these at a Le Studio spring performance, my husband asked, 'Is that as hard as I think it is?' It is.] Well, if they had to do those – they wouldn't. It'd be illegal or something. But if they *had* to, it would be – there would be dancer *baseball cards*. They'd be, like, these *stars*.

This is not an aberrant view. In her essay, 'Pointe Shoe Confidential,' Fisher quotes a 'sore-footed 15-year-old' who observes, 'I don't know who invented these shoes but I'm sure it was a man' (1999: 64). Historical inaccuracy aside, part of the connection that holds these young women together involves gendering their pain and managing this paradox: the ballerina must operate with a high degree of strength and intentionality, and expend considerable energy, to appear delicate and seemingly necessitate male support. Or, put another way, we can borrow from Mary Anne Doane's discussion of the ethereal ballerina's evil twin, the femme fatale: the dancer has power only despite her image, seemingly despite herself (1991: 3).

Young women at Le Studio are atypical in one regard: the large number of men in the company gives them a clearer sense of the gender politics of ballet than that available to most dancers at comparable schools with a dearth of men and boys in class and in performance. Further, they are well aware that this critical mass of 'guys' gives them unique opportunities, both interpersonally and artistically. 'Lisa' observed: 'A couple of my friends came to watch company class one night and they were, like, "Ooo – you get to dance with *him*? And *him*? They are *so cute!*" And I'm like, "Look, once you smell a bunch of guys for three hours in rehearsal its – you don't care what they look like. They might as well be your brothers." ' While there certainly have been flirtations over the years, the Fullers prohibit any non-ballet contact between the underage girls and older men in the company. There are, of course, routine crushes between girls and boys closer in age, but these quickly dissolve in the decidedly unromantic routines of training, as 'Gina' attests: 'My high school friends are, like, "So, do you go out with that guy [her partner in the Arabian variation of *The Nutcracker*]?" And I'm, like, "PLEASE. We're sweaty. We're tired. We're pissy. NO!" Ballet's made me a lot more – *real* about guys. Not like a lot of my friends. Plus it's like I've got all these big brothers. In ballet, it's all out there, believe me, and it's not, like, [air quotes] romantic.'

If these young women see their male colleagues as big brothers, they are also aware of them as rivals for audience attention and regard. Opportunities for partnering work are weighed against the acclaim 'the guys' receive for, in the girls' view, their greater visibility by virtue of their status as relative anomalies. The following narrative comes from a group of three girls. They are very animated; the words tumble over one another so that it is virtually impossible to identify individual speakers. It is almost a kind of chorus:

> The guys have it so easy – not just in performances but in class too. You know, they don't have to wear pointe shoes, that's the big thing, so they don't have all this pain. And they don't have a clue! And they can be older and fat – well, not really fat but, you know, a [professional] company will still take them if they're big, bigger. And they don't have to be as good. Like [a young man in his early twenties who trained briefly at Le Studio] – he was a jerk and he was just, like, *bad* and I heard he got a company contract somewhere. And they don't get dropped and hauled around. And they get all this attention: 'Oooo! *Guys* in *ballet*! How *special*.' It sucks.

This same critical gender consciousness, this awareness of the relative erasure of their own labor when in front of an audience, extends to these girls' readings of audiences' responses to *pas de deux*, the male-female partnerings in ballet. Hanna notes, 'The outstanding and widely recognized sign of sexuality in ballet is the heterosexual *pas de deux* and partnering style in which the man supports, manipulates, and often conquers the woman. Love, with its smell and sight of partners at close range, is signified in the concretization of the embrace, and prolonged ecstasy, through the adjustment of weight and sustained balance. Today the *pas de deux* is often a metaphor for the American ideal of romantic love, a romanticization of romance' (1988: 166). Leaving aside the decidedly unromantic mechanics of producing *pas de deux*, like 'smelling a bunch of guys for three hours in rehearsal,' Hanna's description elides the actual corporeal responsibility for this 'adjustment of weight and sustained balance.' This is something older men in the company, including 'Don,' who danced professionally and later took on character roles in Le Studio performances, knew very well: 'It looks like the guy's doing all the work but, I'll tell you, it's all in how the girl places herself and holds her body. It's a lot easier to lift a good [well-trained] 120 pound girl than a sloppy [poorly trained] 100 pound girl.' Indeed, the girls are advised, in *pas de deux* class, to turn, balance, and execute their steps regardless of what their partners are doing even when, at least nominally, those partners should be holding or supporting them. They are well aware of this gendered labor differential and are quick to name it when they feel audiences are excessively attentive to 'the guys.'

> *'Lisa'*: [mimicking the audience] 'Ooo! Look! He's holding her up! He's working sooo hard! And she gets to stick her leg out and just be carried around while he holds her up!' Nobody thinks, well, he didn't have to get up there and hold yourself exactly so he could grab you – or grab you, you know, *wherever*. He doesn't take off and hope somebody's under there when you come down and he's not gonna fall or have his costume hiked up to *here* and he's not gonna get dropped.

Call this, in response to Hanna's formulation above, the critical de-romanticization of the labor of 'romance.'

McRobbie notes that girls' investments in popular culture fantasies of dance offer possibilities 'of moving into a more independent space which carries with it the promise of achievement while simultaneously holding at bay the more adolescent dynamics of sexual success where a whole other set of competencies come into play' (1997: 230). The words of the young women at Le Studio suggest that actual participation in ballet, at least in this context, offers something more complicated. Here, they gain a solidarity forged through cooperative competition ('you work together so everybody gets better') and group management of pain, in contrast to the 'rugged individualism' Brumberg sees as isolating and pernicious. Further, they share a very real, critical awareness of the powerful and pervasive links between gendered labor and (in)visibility, an awareness and a corporeal confidence forged in community that they add to the 'whole other set of competencies' McRobbie alludes to. As 'Gina' suggests, these young women can, at least in ballet, read and theorize the difference between being 'real about guys' and being 'romantic' in ways their 'high school friends' cannot. They understand the gendered dynamics that diminish their own prowess in the eyes of an uncomprehending audience who might shift this prowess elsewhere.

Yet there is more than a sorority of physical and critical discomfort at work in young women's commitments to ballet, more, even, than the 'simple promise of achievement' in a relatively 'independent space' that McRobbie describes. These young women revel in the pleasures of effective creation as surely as they share its pains. Indeed, as Scarry suggests, such pleasures and pains may be intimately linked. In *The Body In Pain* (1985), she suggests that the interrelationship of pain and imagination, the former discursively objectless and the latter wholly object, facilitates the making and unmaking of the world. Scarry is writing specifically about the pain of torture, but her observations resonate in the infinitely less fraught context of ballet. Consider my conversation with 'Joanna':

> 'Joanna': What's best are the performances. Actually, they're best and worst. The best and the worst. Do you know what I'm saying?
>
> JH: Yes. You feel the best when you feel the worst.
>
> 'Joanna': Yea. Yea. Like in 'Snow' [a variation in The *Nutcracker*]: those *relevé passés* kill you. They're fast and you're tired by that time, and hot, and you hurt and you feel like shit and you have a cold probably, but it's like 'I'm here! I am here!' What they [the audience] see are the moves. They don't know what it's *like* really.
>
> JH: They don't know how hard it actually is.
>
> 'Joanna': Right. But they know *they* can't do it and you look, you know, you sort of look at yourself doing something so hard it's – for most people it's so hard it's kind of ridiculous – and you say to yourself, 'It's me doing this!' It changes what you even think is possible like, in the world. Really. It's like opening up and finding – or, like, making – something amazing and *you* did it! And it's really beautiful. I guess that's really why we all do it. I love it! It's great!

This is the 'gift' of performance that Erica spoke of: an 'opening up' of self-awareness and self-regard ('It's *me* doing this!'), finding, or creating, 'something amazing' from love and labor. Coming from 'Joanna,' these observations seemed especially compelling; she was a bit older than average for girls in the company, though she did not seem so, and from a working-class background. She was one of the few girls whose home life was very difficult; members of the community spoke in hushed tones about parental alcoholism and abuse. Dance, and Le Studio, was clearly a refuge, a safe harbor, for her in ways even more literal than for many of the others. Yet the love she professed for ballet and for performance, and her testament to its power to refresh and renew, was something I heard over and over from these young women.

'Joanna' notes that she may 'feel like shit,' but she does not indulge in 'the hyperbole of conquest and breakthrough' that, David Morris notes, so often accompanies the coupling of work and pain (1991: 23). Her performance is not a matter of conquest but of hard-earned expertise ('they know *they* can't do it'). The audience does not exist to be placated or pleased; at best they are imperfect witnesses, a counter-community registering the separation of those in the know from those who 'don't know what it's *really* like.' Most of all, this 'gift' of technique, forged in the home place of the studio and materialized on stage, 'changes what you even think is possible in the world.'

'Joanna's' comments recall Scarry's observations about pain, work, and creation. She notes: 'The *aversive intensity* of pain becomes in work *controlled discomfort*. So, too, imagining achieves a moderated form in the material and verbal [and corporeal] artifacts that are the objects of the work. . . . While imagining may entail a revolution of the entire order of things, the eclipse of the given by a *total reinvention of the world*, an artifact is *a fragment of world alteration*' (1985: 171; emphasis in original). Scarry writes about 'artifacts,' but it is clear from 'Joanna's' account that the transmutation of 'feeling like shit' into something 'amazing' is also a fragment of world alteration. More than this, her proclamation 'I love it! It's great!' testifies to Joanna's profound 'pleasure in knowing the world can be remade' (Sippl, 1994b) in ways that are self- and community-affirming and, and because, these ways are beautiful ('that's really why we do it').

In a later work, *On Beauty and Being Just*, Scarry explores beauty's capacity to generate pleasure in knowing the world can be remade. Though I am skeptical about some of her ultimate conclusions, aspects of her argument, coupled with her observations about pain, work and creation, offer a way of thinking through young women's commitments to ballet. Beauty, Scarry argues, functions like a contract or compact (1999: 90); she suggests that this contract links the beautiful being to the perceiver. While this might certainly be true of audiences' responses to the ballet as beautiful being, I suggest that this contractual function works somewhat differently in dancing communities who come together to willingly labor in daily service of beauty. Consider the dancer, above, who observed of pointe shoes that 'we're all in it together... and it's beautiful in the end – hopefully.' This 'hopefully,' coupled with Joanna's statement that ballet is 'really beautiful. I guess that's really why we all do it,' suggest another way to see beauty as a contract. As Scarry suggests, beauty incites 'an impulse toward begetting' (1999: 9), a generative impulse, an invitation to participate. Like technique itself, beauty is relational infrastructure. It is also a wish: 'it's beautiful in the end – hopefully.' Beauty in community is the productive possibility of what might be, produced through diligent work at what is. This work is both utterly material, as blistered feet will attest, and utopian, changing 'what you even think is possible in the world.' This coupling of both the material and the utopian is at the heart of the contract binding young women to ballet technique, to Le Studio and to each other. Utopia is, paradoxically, no place. Jill Dolan suggests that utopias in performance are local; these 'imaginative territories that map themselves over the real' become incarnate 'in the interstices of present interactions, in glancing moments of possibly better ways to be together as human beings' (2001b: 457).

The young women at Le Studio have actively, thoughtfully fashioned utopian moments and places, 'better ways to be together' as girls and as performers from a technique that aspires to transcend even gravity. But transcendence is not the path they choose. Instead, they are planted firmly on the point where vernacular landscapes become art, surveillance becomes provisional freedom, individual effort merges with *esprit de corps*, and pain and hard work create beauty that changes what one thinks is possible in the world. From the safe harbors of technique, Le Studio and sorority, these young women also discover how powerful and pervasive – and sometimes how malleable – class and gender relations are with respect to individual desires and motivations, how deeply these are etched onto bodies, and how bodies, individually and in community, both serve and resist them.

PART III HOME BOYS ON FAULT LINES: MASCULINITY AT LE STUDIO

The technical protocols of ballet simultaneously solidify heteronormativity in the classical repertoire and unsettle it through popular stereotypes of male dancers as gay. Part III examines this paradox as it plays out among many of the men and boys who train at Le Studio. In my years of observing young women and girls, sexuality did not emerge as an issue; if there was same sex desire, it was not spoken, even as rumor or gossip. But for the men and boys, heteronormativity is both an unexamined privilege and an ongoing point for vigilance. Here, the ambivalent position of ballet itself – both 'as woman' in Balanchine's words and as virile athleticism – is managed through sharp focus on an aggressive, athletic style of dancing rather than through regulation of individuals' personal sexualities. The complex dynamics between these men and boys sheds light on the fault lines – the actual and potential breaks, cracks and fissures – of heteronormative masculinity as exposed in ballet at the most local level of its production. Heteronormativity exposes ambivalences about the male dancing body, particularly in ballet, even as ballet exposes fractures in heteronormative privilege.

Judith Butler argues that, in the process of subject constitution, 'The power imposed upon one is the power that animates one's emergence, and there appears to be no escaping this ambivalence. Indeed, there appears to be no "one" without ambivalence . . .' (1997: 198). She is refer-ring to 'foundational' subjectivity here but her argument also applies to those arts of existence that build and shape the self accretionally in dancing communities. Young women at Le Studio experience these subjectifying ambivalences condensed into temporal and metaphoric crossroads: intersections of child- and adulthood, individuality and community, home and roam, pleasure and pain. The disciplinary regime of ballet, and the myriad idiosyncratic ways they consume, produce and imagine it, offer them a safe harbor from which to negotiate these cross-roads. This refuge in discipline further offers a sociality of 'both-and' possibilities, rather than only those of 'either-or.'

Like the young women who dance there, the men and boys at Le Studio are also animated by ballet's demands; the 'social categories' of the form 'signify subordination' to technique and generative opportun-ities for existence 'at once' (Butler, 1997: 20). Yet, for this population, ballet and Le Studio are both refuge and, paradoxically, places of tectonic activity where men and boys straddle shifting plates of heteronorm-ative masculinity that threaten to split apart or to collide. For the young women above, subjectification in ballet means that foundational ambi-valences often consolidate into new possibilities. For the diverse group of 'the guys,' these new possibilities are both refuges and sources of anxiety. For Ramsay Burt, male dancers in performance offer the male spectator 'both pleasures and terrors': 'Pleasures may be found in the recognition of sameness and in the reverberation within me of what I learn from my body. Terror lies in the homophobic panic of realizing what these male dancing bodies can do to me and for me, and terrors also lie in having the tables turned on me and being challenged to confront my enjoyment of the voyeurism inherent in my position as a spectator' (1995: 237). Men and boys creating dances at Le Studio manage similar pleasures and terrors – sometimes clumsily, sometimes not – as they reckon with the intimate relationship between dance, community and 'the social pressures to exhibit behaviors that confirm with heterosexual normality' (1995: 237, n.1).

There are several demographic differences between the men's and women's corps at Le Studio. While the women's corps has become increasingly more ethnically diverse over the years, the male corps has always been so: Chicano, Mexican, Filipino, Caucasian, Vietnamese, African American, Japanese, and multiracial. Class affiliations have ranged from upper middle to former wards of the state. At least one company member had contact with the criminal justice system for gang activity. The adults have full-time jobs, all outside of dance with the exception of Xavier, who directs operations at Le Studio Fitness. Full-time employment is especially important to note in light of considerable class, rehearsal and performance commitments.

Michael Peterson notes that 'heterosexual hegemony relies on twin contradictory gestures: assumption and assertion' (1997: 130). In ballet, at least in the popular imaginary, the assumption of male heterosexuality is by no means automatic. For some male students who came to Le Studio, and ballet, for the first time in the late 1980s, assertion had to 'compensate' for the inadequacy of assumption in this context. I suspect some of these aggressive assertions of heterosexuality were also functions of class and culture: whether associated with *machismo* or with presumptions about the 'virility' of individuals' working-class backgrounds. In any case, it was not unusual to hear occasional homophobic remarks from this group at this time. These remarks never referred to, were directed against, or spoken in the presence of, gay dancers and they disappeared fairly quickly, perhaps because they were publicly and consistently linked to the asserters' inadequacies, both as dancers and as members of the community, by other men. In short, whatever the anxieties of other members of the group, these remarks were seen as marking the speakers as outsiders who 'didn't get it,' indeed, akin to the outsiders many company members had to face among their own families. As 'Don' observed, 'this crap' served 'to keep the more insecure guys okay in the dressing rooms. Actually, it's more like they are making themselves okay to wear tights. I've seen it everywhere with [presumably straight] guys in dance. They'll get over it.'

As the number of men and boys taking ballet at all levels increased, gross 'assertion' gave way to other 'self-reassuring stratagems of male heterosexuality' (Peterson, 1997: 123) and homosociality, as well as efforts to manage and negotiate multiple dimensions of difference across the male corps. These centered almost exclusively on presenting ballet technique with maximum visible, and virile, athleticism. Interestingly, these hyper-masculine postures in ballet technique often contrasted, sometimes sharply, with many of the older dancers' negotiations of gender roles outside of dance. A significant percentage was primary caregivers for children or parents. Those with wives or girlfriends had chosen, and were proud of, strong, independent women who figured prominently in the social life of the studio. These 'offstage' interpersonal performances of masculinity had a consistent fluidity that those in the dance studio did not.

Chip and Phip are especially sensitive to their male students, sensitivity rooted, perhaps, in their own experiences. Together with their male students of all ages, they have forged a unique, multi-generational solidarity in a technique whose 'public domain sexuality' is feminine (Hearn, 1992: 200). For the men and boys who take class at Le Studio, particularly those at the advanced levels, time, space, the corporeal chronotopes of home and roam, and the dynamics of effective creation operate very differently than they do for female students.

In some respects, time is kinder to men in ballet. They can begin training later and dance longer. Young women must be extraordinarily focused if they want to pursue dance professionally; late adolescence is the opportune time to decide between college and career. With the exceptions of prodigies like Misty Copeland, girls do not have the luxury of idiosyncratic trajectories into ballet careers, even with smaller companies. Men and boys have much more flexibility, and wider chronological windows, available to them. This gendering of time in technique is reflected in another aspect of Le Studio's demographics. The average age of girls in the most intensive and demanding divisions, Tech II and Advanced, which includes the company, has stayed relatively consistent over the past decade, hovering between 14 and 16. The average age of the men has fluctuated but stays consistently higher: late twenties/early thirties until the mid-1990s and, more recently, late teens to mid-twenties. The actual range has been much broader: from mid-teens to mid-forties at its most extreme.

With notable exceptions, including some who started at Le Studio in Basic I and those who began training at other facilities, the men and boys in the advanced levels entered ballet at relatively advanced ages compared to the girls.[16] Xavier is an interesting example. When he was 12 and living in Mexico City, his mother took him and his older and younger brothers for ballet lessons to, in his words, 'keep us off the street.'[17] He and his brothers eventually decided to audition for places in the prestigious School of the Mexican National Ballet. He and his younger brother made the cut. His older brother, determined to make it as well, came back repeatedly until he too was admitted; he eventually chose a professional ballet career. Xavier came to the United States at 16 and joined Le Studio at 18 after hearing about Phip and Chip. Now in his late thirties, married to a woman who took class in the adult division and with a son and daughter enrolled at Le Studio, he continues to take class and perform with the company. Somewhat more typical is 'Marcos,' a skier, who came to Le Studio as a young adult. He heard it was 'a cool place' and he was attracted by the fact that, unlike many local schools, the hierarchy was predominantly male and 'straight. Not that I have anything against gays or anything.' 'Marcos' came to 'check out' ballet to improve his skiing; he stayed on to take class, become a soloist with Le Studio's resident company and, after seven years, left to take a full-time position as a principal with the Sacramento Ballet before eventually retiring from dance. 'Giovanni' represents a different, less typical trajectory; particularly deliberate about his dancing, he moved from Chicago to study with Le Studio faculty as part of his professional dance development.

If 'Giovanni' is especially focused, he is not unique in his ambitions. While younger members of the male corps occasionally replicate the career trajectories of many of the girls, choosing college or career over ballet once they reach young adulthood, many more have professional dance ambitions or desires. For the younger men, and even some who are older, there are opportunities for scholarships to summer programs and openings with semi-professional and professional companies. In general, even though men dance until more advanced ages than do women, if one is not working in a professional capacity by 25, it is fairly certain that career options for anything other than character work are quite limited.

Many of the older men at Le Studio would choose professional careers if they could but they realize that time has run out. In 'Felipe's' words, they started 'too damn late, except for character stuff. It's too damn late for the prince.' Some go on to pursue careers in dance other than those involving performance. 'Gerardo,' for example, organized a series of dance training and performance workshops for area schools. Others leave entirely; Erica's brother Pablo, who can 'still do double tours in front of [her] face,' followed in his father's footsteps, becoming a barber and raising a family. Likewise, 'Felipe' left to become highly successful in the construction industry, which he entered through contacts made at Le Studio, and is raising a family as a single father. 'Russ,' one of the oldest of the men who trained at Le Studio, chose to 'roam' in an attempt to carve out a professional career in ballet despite his age. Though he won two awards for choreography in regional dance competitions while at Le Studio, he was mindful of the challenges he faced: "What I want is to produce work [choreography and performance]. I've got a couple of commissions and I'm going to Pittsburgh [an intensive summer program for performance and choreography at the Pittsburgh Ballet Theatre] but I've got to drop everything to do it. I've got to do this now, man, and I'm just thinking, 'How can I do this when I have to paint houses to eat?'" Eventually, 'Russ' secured a student position with the Pittsburgh Ballet Theatre. His Christmas letter to the studio, addressed to 'Bun heads, Bun watchers and directors of Buns' and publicly posted on a studio bulletin board, resonated with 'Felipe's' observation 'too damn late for the prince' and cast light on the rigors of a career in ballet begun later in life: 'I'll bet you are having some fun finishing up "The Cracker of Nuts." So am I but no Princely rolls [*sic*] for me this year, nor Arabian, Waltz, or Spanish. Not even Mouse King... but alas I am performing. I do a real mean rat... 28 times. I am an understudy for Spanish and Arabian. Yaaa Hoo, but even if I could do Spanish or Arabian as well as Baryshnikov there are 17 more members of the company in line ahead of me.'

And yet, relatively large numbers of men and boys do take advanced classes at Le Studio and do audition for and perform in company events in spite of the fact that they may never be professional dancers, in spite of the long hours, the physical demands, and the necessity of juggling multiple commitments. For all of them, to varying degrees and in a variety of ways, their time at Le Studio has been life changing, genuinely initiatory, in ways seemingly more dramatic than it often is for the girls. This life changing aspect is, in some respects, not dissimilar to experiences young men may have in youth or semi-professional athletics. On the other hand, ballet is not baseball.

Time in ballet is gendered in another way at Le Studio. Ballet companies and schools, like universities and baseball teams, are characterized by cohorts. When a particularly colorful group enters, makes itself at home, then leaves, there is always a wistful backward glance from those left behind. This cohort may define 'the good old days'; their characteristics reflect community ideals invoked by 'Remember when?' This 'backward glance' consists, in equal measure, of nostalgia and institutional memory, the boundaries between the two blurring so that it may be hard to distinguish one from the other.

At Le Studio this defining cohort was male, a diverse collection of men who trained at the school and performed with its affiliated company during the mid- to late 1990s. As Erica recalled, 'I wasn't with the company yet but *everybody* in the whole school knew who they were. I think everybody who took ballet in Southern California did. They were stars and they were fierce.' Her older brother was a member of this group.

These men were close and cohesive across multiple axes of difference, including age, ethnicity, class and, in a complicated way, sexuality. They epitomized, in their own minds and in those who remember and invoke them, the camaraderie and values of, and the special status accorded to, men at the school. They shared rooms at summer ballet programs. They got together in the evenings and on weekends to watch boxing at Phip's house. As a group, they basked in the celebrity accorded them after performances at home and on the road in ballet competitions where they were met with the screaming adulation accorded to boy bands. Their collective ethos pervaded the entire school, even the adult division where they occasionally took class. As one woman I trained with observed, 'It's like a frat house with these guys. Actually, it's like *Animal House*.'

As the reference to *Animal House* suggests, this cohort met the stereotypes of men in ballet head on with pranks, backslapping and a profound sense of macho solidarity. In addition, they were explicit in their invocation, demonstration, even parody of 'manly virtues' like, in 'Russ's' words 'getting the job done.' 'Getting the job done' meant visibly and obviously working at dance, drilling difficult moves outside of class and rehearsal, and staging pedagogical interventions – sometimes gentle, sometimes not – when 'a member of the team' seemed to be slacking off. If laziness and whining were (and are) devil terms among the female corps, this group took them as personal affronts and moved, generally collectively, to bring the offending party in line. Though they worked with the company girls, their primary relationships were with one another. The implication here was that, for them, the girls were more or less interchangeable. As Erica noted, when she did join the company as this 'golden age' was waning, she was known as 'Pablita,' a little sister rather than a dancer in her own right. This was quite a departure from, perhaps even an unconscious response to, Balanchine's famous observation which, as much as stereotypes about dancers' sexualities, colored the reception of male dancers: 'Put 16 girls on a stage and it's everybody – the world. But put 16 boys and it's always nobody' (in Leivick, 2000: AR19).

With such a large number of colorful men – so large that they sometimes outnumbered women in partnering class – special performing
opportunities were inevitable. Charles Maple, an alumni of one of Le
Studio's affiliated companies who had gone on to perform with the
American Ballet Theatre and came back to Pasadena to work as a choreographer, set a piece that would become this cohort's signature. *Fall of
the Feet* featured seven male dancers in incredible displays of athleticism
and technique: jumping, lifting, even caricaturing their own machismo
in body building poses. Set to a percussive, highly rhythmic score, it
brought houses down. 'Fans' would rush the dancers after performances.
Men and boys at Le Studio still hear tales of this piece, part of a 'golden
age' when, whatever the identities they put into play individually, on
stage 'men were men.' Years later, Maple would choreograph *Eight* for a
new generation of Le Studio 'guys.'

Ballet is particularly amenable to nostalgia, with its reliance on a
classic repertoire that is itself concerned with a romanticized past.
Nostalgia has a bad name; consider Susan Stewart's definition of it as a
'social disease,' a 'transcendent which erases the productive possibilities
of understanding through time' (1993: 60). Linda Hutcheon asserts that
nostalgia 'connotes evasion of the present' (1988: 39). Yet, I would argue,
there are other more generative ways of thinking about the contested
zone between institutional memory and nostalgia. Debbora Battaglia
observes:

> nostalgia does not invariably entail false contact that subverts
> 'authentic' engagement; it is not . . . merely a yearning for some real
> or authentic thing. Rather, it generates a sense of productive engage-
> ment which is at once more personal and larger than any product
> it might find as its object. As cultural practice, then, it abides in a
> convergence of mimesis and poesis – in acts of replicating the social
> conditions of and for feeling, such that one's experience of social life
> is supplemented and qualitatively altered.   (1995: 93)

As Battaglia suggests, nostalgic reflections on these 'good old days'
are lyrical and constructive; at Le Studio, they consolidate a masculine experience of, and ideals in, ballet and communicate these across
generations. Invocations of 'the men's piece,' or other images from this
'golden age' do not evade exigencies of the present; indeed, for individuals, they often serve to highlight them. For Xavier, participation in
this cohort of men marked a time of picking and choosing among prime
roles like those in the Arabian or Spanish variations of *The Nutcracker*

'Russ' mentioned in his letter. He was certainly mindful of this as he trained for a recent version, where he danced only the Russian variation and the Bear. 'There and then' offered a way of reckoning with a relatively diminished 'here and now' as Xavier's age and family responsibilities, and a new, younger cohort of men and boys, shared the stage with his memories of how it used to be. Yet it is also true that his nostalgic reminiscences signaled continuity between past and present; year after year, theatre after theatre, *Nutcracker* after *Nutcracker*, Xavier has danced 'the Bear' in Act I. 'When I retire,' he mused, 'they're going to have to hang up that bear suit.' Here, in moments when over a decade of *Nutcrackers* both run together and fracture along the fault lines of age, '(the hurry [of the present] is stopped) (and held) (but not extinguished) (no)' (Graham, 1995: 156).

This tactical nostalgia is also prospective, offering newcomers the possibility of a similar homosocial community forged through their own labors. Here, memories of this 'golden age of guys' functions less as a backward glance than as what Morson calls 'sideshadowing' (1994: 117). Morson suggests that sideshadowing offers a 'middle way' of experiencing time and possibility. Here, it exists between the realities of Le Studio's emblematic cohort (*Fall of the Feet*, its exemplary execution and the responses it generated), and its impossibilities (assertions that this camaraderie was perfectly seamless and unproblematic). It also, simultaneously, points to 'the impulse to the hypothetical' (Morson, 1994: 6, 148). Thus, Charles Maple's recent piece, *Eight*, recalled the athleticism of *Fall of the Feet*, acknowledging the differences in technical facility between 'then' and 'now' by including boys in the dance, and offered the possibility of a new community in ballet, organized into a reassuring spectacle of masculine agency and power.

Like time, space at the school is also gendered. For the young women who train there, ballet generally and Le Studio in particular are congruent home places within the larger context of Los Angeles. In some ways, they are interchangeable. For men and boys at the advanced levels, they are not. Further, while girls use this home place as a way to hide in the light from pressures of parents and 'teenage stuff,' men and boys use Le Studio to hide, ironically enough, from ballet itself, or at least from the stereotypes that accrue to it. Le Studio becomes a unique safe harbor, offering its male constituents the pleasure of dance while keeping its 'perils' at bay.

The men and boys who train at Le Studio and dance with its resident company are special and they know it. 'Russ's' comments are representative: 'What makes [Le Studio] special are the guys. Most companies would kill to have us. You just have to say that. Good girls [baller-inas] are easy to find but guys – that's why we're the show. We're it.' Part of this self-confidence arises from a shortage of male dancers in amateur and semi-professional ballet companies. It is quite unusual to have large numbers of men and boys in training at a neighborhood, or even a regional, studio in the United States, outside of, perhaps, New York. Further, Le Studio's male dancers are noted for the high quality of their technique and for a style that one company member character-ized as 'take no prisoners.' Yet this sense of being special and unique must be leveraged against a larger cultural unease about male dancers. As Ramsay Burt notes: 'for many people, a key source of contemporary prejudice is the association between male dancers and homosexuality. It is certainly true that there are a lot of gay men involved in the dance world. Although by no means all male dancers are gay, this is what the prejudice suggests. One explanation of macho display dance is some-times surely that dancers are trying to show that they are not effeminate, where "effeminate" is a code word for homosexual' (1995: 12). Recall that Erica's brother categorically asserted there was 'no way' he would take ballet classes because 'that's for boys who are gay.'

'Take no prisoners' technique is shorthand for a specific performance of heterosexual masculinity that is loosely defined and nonprescriptive, but nevertheless pervasive in men's and boys' training at Le Studio. Indeed, as with 'Marcos' above, this is what many male students say attracts them to the school. Yet ballet's associations with effeminacy and/as gay sexuality means that even 'take no prisoners' performances do not sufficiently reassure some students. Some seem defensive, both about their choice to take ballet at all and about their own presumptions and negotiations of ballet and sexuality in this community.

'Brett,' who began to train there in his early twenties, observed: 'I just feel more comfortable here. I'm not a bigot or anything. I'm just – more comfortable. Like a woman who wants a woman doctor because men, you know, can't relate. There's nothing wrong with that.' 'Brett's' analogy and his seeming defensiveness are revealing. For him, ballet training demanded the same vulnerability, and required the same level of trust, one would expect from a health care professional who must be able to 'relate,' presumably on the grounds of bio-social similarity. Sexual difference here elides unproblematically into gender difference: fissures in masculinity are equivalent to a perceived gulf between women and men in matters of 'comfort' with physical exposure. Yet, for all his invocations of 'feeling comfortable,' 'Brett' was also clearly uneasy about his own heteronormative presumptions. His tone in this observation was tense and brittle, particularly in his assertion that he was 'not a bigot.' This unease recalled 'Marcos's' qualification, 'not that I have anything against gays or anything.' Perhaps this defensiveness arose from 'Brett's' awareness that multiple dimensions of male difference, including sexuality, were currently at home in Le Studio, possibly within earshot of his statement. Perhaps, as a young multiracial man who came to dance from the visual arts, 'Brett' was self-reflexive enough to recognize the possibility of an underlying homophobia when he heard it, even from himself, or perhaps he resented being asked to account for a 'comfortable' choice to which he felt entitled. In any case, his uneasy assertions of his comfort suggest a deeper ambivalence undergirding the place of straight men in amateur ballet and, in a larger sense, in the structure of heteronormative masculinity as well.

As the 'male members' of an activity widely coded and read as feminine/effeminate, the men and boys at Le Studio are in the paradoxical position of constructing and occupying a place that exposes fissures in dominant, totalizing fictions of masculinity even as it provides '*resources* for men to gain, perhaps regain, their power and their sense of being a man' (Hearn, 1992: 20; emphasis in the original). In ballet generally, in class and in performance, males at Le Studio inhabit all the contradictions intrinsic to the patriarchal, heteronormative construction of 'public man.' Hearn notes:

> When I refer to 'public men', I am thinking not primarily of 'public figures' or individual 'men in the public eye', but rather of different men's presence in and relationship to the public domains. The notion of public men as 'public figures' or individual men 'in the public eye' is itself an ideological elaboration of the general construction of men in public. Their individual power accrues from the general, that is a social structural, relation of men to women, in the public domains and elsewhere. 'The eye' in 'the public eye' is not an individual eye but a structural arrangement, the social structuring of visibility and invisibility, to which all may be subject. In another sense, the notion of public men in use here is itself ideological and could be contrasted with the notion of private man as the apparently autonomous patriarch. (1992: 3)

Hearn argues that the experience of public men is embedded in praxis that links 'experience, work, politics, theory as human, sensuous activity' (1992: 22). Further, this experience is fraught with 'diversifications and fragmentations' that subdivide public patriarchies as well as types of masculinities'; 'disjunctions and fracturings' that permeate experience; contradictions in the social processes supporting public masculinity; and 'the persistence of the consolidation and unities of men as a gender class' (Hearn, 1992: 22). All of these ideological operations are managed and deployed by men and boys at Le Studio as their own gendered arts of existence; sometimes these arts reassure and regulate through their consolidating force. Sometimes they exceed or fall short of generating soothing fictions of coherent masculinity, revealing seismic fragmentation and fractures. Sometimes they do all of this at once.

The ambivalent position of public men in ballet has not escaped the notice of popular culture. Indeed, popular culture representations often focus on very little else when representing men in ballet. The recent film *Billy Elliot* offers an example of ballet as a consolidating activity; it reconciles traditional working-class masculinity with ballet by using chronotopes of home and roam and the motif of the unlikely context to support a story of male family fracture and, ultimately, coherence. The film apparently struck a chord. Junior and senior schools of London's Royal Ballet saw a marked increase in boys' enrollment after the film's release. In the junior division, more boys than girls were admitted in 2000 ('Morning Report'). Paul Thrussell, a principal dancer with the Boston Ballet, narrated his own autobiography in terms of the film in the *Boston Sunday Globe* (2001).

Billy Elliot may have succeeded in temporarily suturing the fissures in patriarchal heteronormative masculinity exposed by ballet, but another earlier popular culture text more accurately reflects American ambivalence toward male dancers. Here, Bart Simpson, 'this century's Denis the Menace,' finds himself enrolled in ballet after all other options for physical education are filled (*The Simpsons*, 'Homer vs. Patty and Selma' 2F14). He enters the class as the teacher, a woman with a Russian accent, tells a room full of giggling pink-clad girls: 'Today we learn the Dance of the Fairy Queen. You can either be a fairy or a queen.' Proclaiming that 'ballet is for sissies,' Bart turns to leave. 'Ballet is for the strong,' the teacher insists, 'for the fierce, the determined, but for sissies? Never!' Then she adds, assigning him his part, 'You are a fairy.'

Bart's conversion happens as he struggles out of his unitard, contorting himself into splits, *pirouettes, grand jetés*: 'Wait! Joy of movement increasing! Love of dance impossible to resist!' The girls gasp in admiration. Next we see Bart in class, a purple head scarf matching his purple unitard. He tosses off a combination and observes coolly, 'See that? I started to do, like, a little *arabesque*, but then I just fully went for it and pulled off the *demi entrechat*.' He pauses to flip open a can of TAB, adding, 'Not that I'm into that kind of thing.'

The crisis arrives in the form of a school recital ("Ballet Performance: The 'T' is silent."), attended mostly by students in detention. Principal Skinner observes, 'We even bussed in trouble makers from other schools.' Bart's teacher is sympathetic: 'I know you have great conflict. You love ballet but you fear the boys will laugh at you, no?' 'No,' Bart replies. 'I fear the girls will laugh at me. I fear the boys will beat the living snot out of me.' Yet he takes the stage, his face concealed by a mask. His school peers are enthralled, particularly delinquents Jimbo and Nelson. Jimbo observes, 'He's graceful, yet masculine, so it's okay for me to enjoy this,' adding, while dabbing his eyes, 'I haven't been this moved since *The Joy Luck Club*.' Rapturous applause convinces Bart to reveal his identity: 'Go ahead and laugh but I took a chance and did something I wanted to do and if that makes me a sissy then I guess I'm a sissy!' Jimbo, now dry-eyed, yells to his crew, 'Bart's a sissy! Let's rush him!'

Chased to the edge of a chasm, Bart attempts to escape by leaping across in *grand jeté*; his ballet teacher, appearing like Obi Wan Kenobi, intones, 'Bart, use the ballet!' But the force is not with him and he crashes to the ground. 'Looks like he took a pretty bad spill,' observes one of his pursuers. 'Well,' says Nelson, poking Bart with a stick, 'as long as he's hurt.' His sister Lisa arrives and, praising his sensitivity, proclaims him her kindred spirit. We are left with a contorted Bart, still on the ground, wondering aloud, 'Why did she leave me here when I need medical attention?'

The episode glosses the multiple contradictions of men in ballet in the popular imaginary in ways that supplement the comparatively simplistic politics of *Billy Elliot*, where dedication to technique and natural talent ultimately win out over all. Here, ballet is for the fierce, yet Bart is assigned to play 'a fairy.' The only other option is playing 'a queen.' The spectacle of 'that masked pointy-haired male dancer' is moving, but public ownership of that spectacle means you get the snot beaten out of you. The grace and athleticism of ballet don't come through when it counts and the contortions Bart is left in after fleeing his would-be tormentors seem a parody of his moves on stage. Even his sensitive sister, the resident intellectual Lisa, ultimately abandons him to his misery after hollow praise. 'Home' is not a refuge and 'roam' is not an option. The narrative leaves us with a sense that we can admire Bart for doing something he wanted to do – as long as he's hurt for affronting conventional masculinity. Interestingly, in this episode of *The Simpsons*, the ballet plot is secondary to the main narrative arc in which Homer must debase himself before, and ultimately redeem his masculinity from, his shrewish sisters-in-law to whom he is in debt.

Yet, though the narrative discursively punishes Bart for trying ballet, it also reveals some uneasy elements of male spectatorship in Jimbo's sniffles, his reference to *The Joy Luck Club*, a 'chick flick,' and his observation that Bart's self-evident masculinity in movement makes it okay to enjoy his performance onstage. Bart has to be punished, both within the narrative and by the narrative, not just for potentially being a 'sissy' but because of Jimbo's response to watching him. Here, the male dancer signifies a profoundly gendered ambivalence: his masculinity consolidates his relationship with other men through displays of physical prowess but also fractures that solidarity by suggesting difference and exposing potentially dangerous pleasures of same sex spectatorship, pleasures that, in themselves, threaten to fracture illusions of a universal, coherent masculinity still further. This is a graphic illustration of the pleasures and terrors of the male dancing body Burt described above.

These ambivalences are, in part, historical, rooted in the shifting status of men in ballet over time. As Judith Lynn Hanna observes, men dominated the form until roughly the Romantic era (1988: 128–9). Garafola notes that, '[B]y the late 1840s, the handwriting was on the wall for leading men' (1997: 7), many of whom stopped performing entirely.[18] Likewise, boys' enrollments in ballet classes plummeted precipitously. Ramsay Burt argues that this decline in the status of male dancers was political and directly attributable to a rising middle class's binarism of disgust.

> Prudish attitudes toward ballet and the male dancer in particular were surely another area in which middle-class distaste for what they perceived as aristocratic degeneracy was expressed.... While on the one hand middle-class sensibilities were disgusted by the specter of a degenerate and decadent aristocracy, they also feared and were disgusted by the vigour and fecundity of the working class.... For the mid-nineteenth century bourgeois ballet critic, vigorous and manly displays of dancing might sometimes have carried negative connotations of working-class entertainment.   (1995: 26–7)

Yet, as Ann Daly argues, the pendulum would eventually swing in the opposite direction: 'Male dancing rose to prominence during the [nineteen] seventies – at the same time, ironically, that the women's rights movement reached its peak. The shift was accompanied by a lot of "dancing is masculine" propaganda in the press... and in a spate of books,' (1997: 115), as well as by the extensive marketing of male ballet 'stars.'

As Daly's observation suggests, class bias masks a more primary unease about the gender of dance itself. Two utterances from the two temporal poles of male dancers' status are especially illustrative. The first comes from nineteenth-century critic Jules Janin, who conflates gender and class bias in his disdain for men in ballet:

Speak to us of a pretty dancing girl who displays the grace of her features and the elegance of her figure, who reveals so fleetingly all the treasures of her beauty... But a man, a frightful man, as ugly as you and I, a wretched fellow who leaps about without knowing why, a creature specially made to carry a musket and a sword and to wear a military uniform. That this fellow should dance as a woman does – impossible! That this be-whiskered individual who is a pillar of the community, an elector, a municipal councillor, a man whose business it is to make and above all unmake laws, should come before us in a tunic of sky-blue satin, his head covered with a waving plume amorously caressing his cheek, a frightful *danseuse* of the male sex, come to pirouette in the best place while the pretty ballet girls stand respectfully at a distance – this was surely impossible and intolerable, and we have done well to remove such great artists from our pleasures.
(In Daly, 1997: 113; see also Burt, 1995: 25)

Almost 140 years later, the image of the bourgeois public man is no longer the male dancer's social foil; now, as Daly argues, the ballerina is. She quotes *New York Post* critic Clive Barnes's assertion, headlined 'How Men Have Come to Rule Ballet's Roost': ' "Male dancing is much more exciting than female dancing. It has more vigor, more obvious power, and an entirely more energetic brilliance. Of course there are different qualities – thank Heaven! – to female dancing, yet there is something about the male solo, its combination of sheer athleticism with art, that makes it unforgettable." ... Female dancing, he implied, is valuable *only* because it is different; the important – and "unforget-table" – qualities are already and exclusively embodied in male dancing' (1997; 115). Burt observes, ironically, that the rhetoric of 'energetic brilliance' surrounding artists like Rudolph Nureyev and Mikhail Bary-shnikov was rooted in presumptions from the very era in which men in ballet were particularly disdained: 'The Romantic idea of male artistic self-expression clearly underlies much of the hype that has surrounded the recent popularity of the male dancer.... As far as theatre dance is concerned, during the nineteenth century the dancing of ballet movements was not recognized as a reputable means of artistic self-expression, let alone a means through which male genius manifests itself' (1997; 22).[19]

Barnes's emphasis on 'sheer athleticism' suggests how masculinity in ballet could be rehabilitated and further, how it could exist along-side the famous and roughly contemporary, oft-quoted proclamation of George Balanchine that 'ballet is woman.' Bravura jumps, acrobatic feats of agility seemingly effortlessly executed, and a more aggressive style of partnering affirmed the form's masculinity onstage through a seeming choreographic dominance over women dancers and forged a link between male dancers and male athletes (see Daly, 1997: 11–12; Burt, 1995: 5–6; Hanna, 1988: 143–6). Indeed, ballet dancer Edward Villela merged the twin signifiers of masculinity, fiscal and physical prowess, by favorably comparing his salary to that of his sports contemporary, pitcher Tom Seaver (Hanna, 1988: 145).[20] This rehabilitation through public proclamation of athleticism and/as virility was heavily marked, and marketed, in ballet, even more so than in modern dance which, while founded and championed by strong women, was also historically somewhat more hospitable to men (see Burt, 1995: 101–34; Carman, 2001). These invocations, and comparisons with sports figures, served as rhetorics for consolidating public masculinity.

This rhetorical conflation of ballet and athletics was especially important; if the classing and gendering of dance generated what Burt called the disconcerting 'trouble with the male dancer' (1995: 10), this unease was also undergirded by an even more foundational ambivalence, one implicit in *The Simpsons* episode described above: the gendering of spectacle. Burt argues that, over the past 200 years, 'it is not that male dancers have quietly absented themselves but that they have been nervously dismissed' (1995: 13), along with the male nude in painting and sculpture. These comrades in spectacle presented the 'danger of infringing the conventions which circumscribed the way men could be looked at' (1995: 22).

Various models of gendered spectatorship and identification have been offered to account for what John Berger has labeled 'looked-at-ness' in ballet. Laura Mulvey's Lacanian reading of Hollywood film is particularly popular (see, for example, Briginshaw, 1997; Burt, 1995; Koritz, 1995; Wolff, 1997). While the approach does have its critics among dance scholars (see Thomas in Morris, 1996; Banes, 1998), Ann Daly has persuasively argued that Mulvey's 'male gaze ... remains a fundamental concept: that, in modern western societies the one who sees and the one who is seen are gendered positions' 1997: 125). As Burt argues, historically: 'not only did the male dancer become out of place on the newly feminine ballet stage but, because male appreciation of the spectacle of ballet took on sexual aspects, the ways that male dancers appeared on stage became a source of anxiety to bourgeois male spectators. To enjoy the spectacle of male dancing is to be interested in men' (1995: 28). That is, viewing ballet was a consolidating erotic spectacle confirming patriarchal privilege when/as men looked at ballerinas; the male dancer introduced the possibility of recoding heteronormative viewing, fragmenting this solidarity of seeing. To be 'interested in men' is to disrupt the homosocial contract by raising the possibility of homoerotics in spectatorship. By emphasizing athleticism, ballet alleviates the male heteronormative anxiety of being 'interested in men.' Bravura physicality reassures male viewers about the source of their own pleasure. As Jimbo observed of Bart Simpson, as long as he was both graceful *and* self-evidently 'masculine,' not simply male: 'It's okay for me to enjoy this.'

Standing out: consolidating homosociality in dancing communities

The diverse group of men and boys who train at Le Studio are well aware of the ambivalence inherent in making spectacles of themselves by choosing ballet. This awareness of their own visibility is one of the consolidating technologies that binds them together in ways very different from those used by the girls who 'hide in the light' from parental pressure and 'teenage stuff.' As 'Marcos' observes, 'There's good news and bad news [about being a man in ballet]. The good news is you're gonna stand out, and that's the bad news. The bad news is you're gonna stand out.' One stands out, not only in class and on stage, but also among those men choosing more conventional physical performances of masculinity including, often, one's own father, brothers, peers and friends.

For most of the young women at Le Studio, solidarity in ballet supplements other social networks ('If you can't talk to your parents about it or you can't talk to your high school friends about it –'); for the men, it meets other socialities head on, sometimes defensively, sometimes offensively. 'Russ' is joking here, but only barely: 'That's why we're [the guys in the company] pretty tight. I mean, can you see it? What are you going to say? "Hi guys. I'm a red-blooded, straight ballet dancer. Good thing I can kick your ass."' 'Russ' is the son of a strict military father, for whom, he says, he is a 'terminal let down'; his observation suggests that, in cases where one might 'stand out' from family expectations, it might be best to hyperbolically assert portions of those expectations ('red-blooded, straight, male'; 'kicking ass') alongside one's ballet ambitions. Further, 'Russ' attests to a specific consolidating performance that links the studio men to each other: 'standing out' by choosing a nontraditional masculine subject position requires, at moments in what he calls 'civilian' life, a kind of defensive compensatory hypermasculinity:

> *'Felipe'*: We go out, you know, and you're wearing a Moiseyev [dance company] shirt and you [slapping Xavier on the back] beat on each other more. You know –
> *Xavier*: You grunt more.
> *'Russ'*: – open a few beers with your teeth. Can't have them drawing conclusions about that dancer shirt, right boys?

These men speak to a vigilant, heteronormative gender poiesis: in the appropriate circumstances, rather than fully inhabit the kind of ambivalences they may inspire, a more 'comfortable' heteronormative position is simultaneously located and performed. We can think of this discursively, using Meaghan Morris's view of 'the movement of the personal': performative hyper- and hetero-masculinity is anticipatory. It is oriented 'futuristically toward the construction of a precise, local and discursive context' (1988: 7) in an attempt to heal fractures in public masculinity even before they are exposed. These public performances of macho behavior recall Erica's observation that she and her friends don't flaunt their ballet personae outside of dance. These men suggest that, if a 'dancer shirt' or other signifier of ballet is deployed to assert identity in public, it is aggressive masculinity that must be 'flaunted' as a kind of reassuring heteronormative antidote.

There are other creative options for managing and extravagantly performing homosociality, this time among 'the guys' themselves. These include trickster turns that call men on their own anxieties, rendering these visible and laughable, particularly to those dancers who are very secure about their places in ballet and enjoy tweaking others who may be less so. Xavier notes: 'Sometimes with the [company] guys and the guys in the Adult classes, we split into groups: two groups of women and one group of guys for center work. We'll line up – just the guys – and I'll say, "Let's hold hands." Or we'll go across the floor together – again, just the men – and we'll be all pulled up and prepared to go across, and I'll put out my hands.' He extends his arms and hands, wiggling his fingers as if beckoning his compatriots to join him in a group sing, then mimes their response: quivering cringes reminiscent of cartoon elephants spooked by, and shrinking from, cartoon mice.

It is important to note that the projections and stereotypes which inspire these performances do not arise only from other men. Women and girls also inspire performative persona adjustments. As Bart Simpson observed, the boys would beat him up and the girls would laugh. Consider the recollection of 'Ms E,' 'Michael's' mother. 'Michael' began training at Le Studio in Basic I, becoming a powerful and expressive soloist with the company and, eventually, a professional in dance.

> 'Ms E': He wanted this, always did, from the start. I never would have suggested it. He was seven. I'll never forget it. You know, they don't make tights for men that young; he had to wear girls' tights and they're thin and everything shows. So we went to class and he went in and all these little girls, they just laughed. They laughed at him. The guys [the Fullers] stopped it, but I could have killed them. I thought, 'You little bitches. How dare you.' When he came out and we got into the car, I didn't want to – I didn't know what to do. I was more upset than he was. He said, 'Mom, next week can we come early? I think if I'm in the studio first, maybe they won't laugh then.'

Support from home can help a young male dancer negotiate the ambivalence of 'standing out' in ballet. Consider 'Jonathan,' age 9 who, after watching his stepmother take adult ballet classes, auditioned for and was cast in the annual *Nutcracker* production: 'They [his stepmother and father] didn't push me but I knew they'd think it was cool. Especially when I do Baby Mouse. Baby Mouse is hard cause it's hot in there [in the costume] and you can't really see out of the eyes, so I have to watch "Felipe" [who dances Mouse King and who is Jonathan's hero] and kind of hold the head on, but they think it's cool and it *is* cool. I like it better than soccer. More people watch.'

Likewise, a lack of parental, and especially paternal, support has consequences. 'Felipe' came to Le Studio after brushes with the law. Ballet changed his life yet, despite his success onstage and off, 'Felipe' was acutely aware that his father had never seen him perform and deeply disapproved of his ballet work. In a moment of extreme vulnerability, he solicited Chip and 'Don,' an older dancer who sometimes performs with the company and from whom he rented a room, as emissaries to plead his case to his father. Despite repeated assertions that he would 'retire' after that year's *Nutcracker*, admittedly an annual proclamation for 'Felipe,' his father, to my knowledge, did not attend any of that year's performances. Though 'Felipe' was a particularly expressive member of the Le Studio community, he did not make this widely known. His only acknowledgement of these troubles at home in my presence came after listening to 'Jonathan,' quoted above, at the opening night reception for the show when he observed, very quietly, 'More might come but more don't always count.'

As these experiences suggest, men's and boys' attempts to integrate their family homes into their homes in ballet at Le Studio are met with more extreme responses than those facing the girls. They are either supported and encouraged or ignored and dismissed. Yet dancers from both extremes are eager to forge a homosocial solidarity with other men at Le Studio and they deploy a wide variety of tactics to construct community. There is a discursive currency of consolidation at work here and its primary coin is 'buddy.' 'Buddy' expresses camaraderie, relative hierarchy and social generosity. It is used to soften criticism, to elicit complicity, to signal belonging.

> *Philip Fuller to 'Michael,' gently chastising him about his weight*: Caesar salads from here out, buddy. Caesar salads.

> *'Felipe' to a younger, less proficient member of the male corps*: No. No. Second [*arabesque*, a pose]. Whadya call that? Is that – what *is* that? Two and a half or something? That's *ugly*. [To Phip] Hey, buddy, wouldja look at what he's doing? What the hell do you call that?

> *'Russ' to 'Jonah,' who is, for reasons discussed below, a 'problematic' member of the male corps*: Buddy, your shoulder is too far back on that [*tour en l'aire*]. It's pulling you back. [He demonstrates.] Straight up and down, buddy. Straight up and down.
> *'Felipe'*: Are you watching him, buddy? Straight up and down.

It's easy to read 'buddy' here as a particle used strategically to avoid the intimacy of naming. Instead it is a connector, an endearment, a gender-specific signifier of relative, or permitted, social parity. The following conversation was conducted partly in jest. The rules described, however, are seriously and scrupulously adhered to:

> *JH*: So not everyone is 'buddy' then?
> *'Russ'*: Nope. Not everyone is buddy.

JH: Is [a former *pas de deux* instructor in his late sixties, former member of the Ballet Russe, deeply respected by the male corps] 'buddy'?

Xavier: He's 'Sir.'

'Felipe': He's 'coach.' ['Felipe' is dissembling a bit. In fact he calls the Fullers 'coach' and would never call this man the same to his face, though 'Don,' who studied with him in Paris, might.]

JH: Is [a former ballet mistress] 'buddy'?

Xavier: No women are buddies!

'Felipe': (almost at the same moment, laughing) No women buddies!

One long-time studio observer, a 'buddy' in his own right, affectionately pointed out that, to this day, the new, younger generation of 'guys,' 10–11-year-olds, perform their status by calling each other 'buddy.'

If there is a physical analog to this exclusively masculine discursive unifier, it is the 'DB' or dance belt, perhaps best defined using Lawrence Ferlinghetti's formulation: 'Women's underwear holds things up; men's underwear holds things down.' Consider:

'Felipe': Hey, coach, forgot my db.

Philip Fuller: Duct tape, buddy. Duct tape.

Hiding sweaty dbs so they stiffen; lewd comparisons of db sizes and conditions; and suggestions of where or how one could or should use, wear or place one's db link men in the advanced levels to one another and to the larger community of men who take ballet at Le Studio. In contrast, girls in the company labeled 'where your eyes go when guys demonstrate' as 'Melvin.'

For the young women in advanced ballet classes at Le Studio pain, and particularly the pain of pointe shoes, is consolidating, shared in complaints and rituals for negotiating sore feet. Pain is also used as a tool to forge solidarity among the men and boys as well. Here, though, the rhetoric surrounding it is frankly, almost hyperbolically, that of hoary formulations of conquest and endurance, the equivalent of walking to school barefoot for ten miles in the snow. Because there is no masculine common agent like the pointe shoe to which all can refer, the extent of the pain conquered or endured must be demonstrated and dramatized lest others have doubts. Such demonstrations might be purely physical: spectacular sweat-soaked flops to the floor after a vigorous class, or knee cracking contests to see whose joints pop the loudest. Or they may be extravagantly verbal, like 'Russ's' comments to his former roommate at the Pittsburgh Ballet Theatre's summer program regarding his new class schedule in a public 'letter home': 'Remember the summer and all those classes? This is just a continuation of the same grueling torture. Add one Nazi Russian [*sic*] dictator of the people's leg mass six times a week and that is all that has changed. The teacher's name is . . . ; avoid her at all costs or suffer and die gurgling your own sweat while standing at the barr[e].' The subtext here, only barely disguised, is that 'Russ' now has it even tougher than he and his roommate did the previous summer.

Competition and closeness in ballet are intimately linked for young women at Le Studio. The men and boys who train there, however, emphasize first one, then the other. They may slap each other on the back when out together in public; they assuredly confide in one another at the ends of long evenings in rehearsal; and they generously coach each other on difficult moves in empty studios after class. In class, though, competition can approach the parodic, particularly in same-sex center and diagonal work: turns and jumps. There have been '*tour*' [*en l'aire*] contests, even 'lift' contests during partnering work. These latter, focusing on who can lift girls higher and faster, gave rise to names like 'the death fish' and 'angel of death' for moves that theatrically dipped or lifted women. The Fullers often use this competitive impulse explicitly to motivate boys in the earlier stages of training. When their mother, in her eighties, visited the school and looked in on classes, Phip admonished the boys in his basic class: 'This is my mother and she can jump higher than you just did.' With men in the advanced levels, no admonishment is necessary as competition takes on a life of its own, prompting Chip to observe, 'There's a lot of testosterone in this room,' and Phip to caution, '*Controlled* aggression, gentlemen.' These cautions are rooted in experience; sometimes 'testosterone' has erupted in unsettling ways, as Phip once observed to me after a rehearsal of *Fall of the Feet*: 'You should have been there today. They just got into it – got *too* into it. It was scary. The counting part ["Felipe" shouts counts to the other dancers in one segment] just, just got out of control.' 'Bob,' who takes adult and advanced classes at Le Studio but is not a company member, added: 'Yah. It was like "Kill the Pig!" [an allusion to *Lord of the Flies*].' Occasionally, these eruptions turn against the dancers themselves. In another rehearsal of *Fall of the Feet*, a fist fight broke out during this same segment.

'Russ' also describes a particularly colorful instance of the competition-community oscillation. As a member of the student division of the Pittsburgh Ballet Theatre (PBT), he brought with him the same 'Animal House' ethos that characterized the 'golden age of men' at Le Studio, only to discover that this *bonhomie* translated very differently in the new context. His narrative, from another public 'letter home,' is so engaging, and so emblematic of the theatrical masculinity shared by his former mates, that I quote it at length.

> *'Russ'*: HEEEEEY, I almost forgot to tell you about one of the privileges of performing as Rat [in *The Nutcracker*] with [the PBT]: if you are new to the company or have made rude remarks about a company member's ability to dance, flagrantly shown off, stolen off with a company member's girlfriend (or boyfriend), inadvertently used a company member's Rat costume (gloves, helmet or tail), or you are just plain stupid and challenge a company member to a double *tour* contest, the consequences are painful. Henceforth. . . .
>
> When you as a new Rat are dressing into your Rat costume and mosey fat and rat-like over to your designated entrance wing for the fight scene, that company member you have pissed off is waiting for you. You will never see him or feel his delicate touch and happy in the belief you are safe on stage you totter out to do battle with the tiny little child soldiers.
>
> Ha! You've been secretly marked; a small bright white piece of tape has been attached to the back of your costume. A small piece of insignificant white on your dark brown fluffy Rat back.
>
> Wham! Suddenly the battle is no longer a choreographed Balanchine masterpiece of toy soldier versus the Rats but all Rats against one Rat, you! In this case, me.
>
> And I was mistaken to believe that my PBT school buddies, on stage with me among the company Rats, would help me fight against the big black-eyed, long plastic-whiskered Company Rats that are now descending, unbeknownst to me, upon my unprotected back.
>
> Har-de-har-har. My school chums are only to be counted on to enthusiastically join in the attack against me lest they become the victim of the Rat Platoon.
>
> The punches come from the bleak blackness like hammers; my legs are slipped out from under me like they were meant to be above my head. This is repeated twice more until finally, my butt bruised and sore, I manage to crawl to my feet only to be clubbed to death by two Rats wielding thick rat tails. Like a thundering Rat tale [*sic*] waterfall they wail their tails against my plastic Rat head.
>
> Out in front, in the audience, it looks like fun: 'The Rats are so animated tonight.'

Meanwhile, punch drunk, my ears pounding from the thundering of Rat tails against my hollow Rat head, I again crawl to my feet. I have now acclimated to the lack of vision from inside the Rat head and, avoiding all attempts to knock me down or pound my Rat head into Rat stew, I quickly begin to pick my targets.

Always attacking from the back with a sweep of the leg and a toss of my arms, Rats begin to fall. Rolling and toppling, the whole Damn Rat Platoon is a fluffy brown rolling furball, Rat legs kicking into the air. I move at speeds even real Rats would be proud of.

The Rat Platoon begins to get back up again; they organize and charge me en masse just as the Rat King is killed. He is killed just in time too, for I was the only Rat left standing and the company Rats have clearly seen which Rat sent them all sprawling to the hard marley floor.

Because the King is dead we, as professionals, have jobs to do. There is dancing to be done, parts to be performed and as professional Rats we accomplish our tasks in a professional manner, crying over the dead Rat King, giving him CPR and carrying his limp, lifeless body from the stage.

As the Rat Platoon waddles back to the costume staging area, no words are exchanged, no threats or hard stares are made as we take off our Rat heads. All is quiet as snow begins to fall softly onto the stage.

Someone's rat ear is hanging, twisted and broken from his Rat head; the seams of two rat tails have been split open and the white cotton stuffing is oozing from the wound. One rat body is ripped partially down the zipper. My left buttock is throbbing horribly and one company member is sure his lip is split and bleeding. He keeps wiping his mouth and looking at the red streaks on his hand. Sometime later in the hallway and in better lighting he discovers it was his lipstick and not his blood he wiped from his mouth.

Everything about the battle that raged earlier on stage is ignored, there is no discussion of who hit who and who got out of hand. The silence is like a thick heavy blanket and there is only the fear of what might happen next time that hangs like a thick choking smoke in the air.

But I will be ready for them next time, then again eight against one aren't the best odds, even in Nutcracker land and especially now that I no longer have the element of surprise. The company Rats know who knocked them down, there will be reprisals, debts to be paid and I have 16 more shows to go. My bottom hurts just thinking about it.

Ah-Ha. I have a great idea and sneak off to steal a roll of white tape. Next performance I will secretly white tape every Rat in sight. Lets [*sic*] do battle baby!

Suddenly a black thought creeps into my brain. All of the company members are issued brown shoes for Rat feet. My shoes are black. Shit.

Hearn observes, 'Men's cultural/organizational experiences were, and are, inevitably complicated and contradictory' (1992: 200). Competition in ballet is one of these. '*Tour*' contests bind men together as audience for each other's athleticism, providing a frame in which it is heteronormatively acceptable to be interested in the spectacular displays of other men. Yet, as 'uncontrolled aggression' and battles with the Rat Platoon demonstrate, this is a fragile ground for community, one always threatening to atomize the very homosocial solidarity it seems to provide.

Anxious pedagogy, ambivalent agency

Competition is not the only fault line running through homosociality at Le Studio. There are two others that problematize masculine agency and challenge solidarity. Both involve difference. Eve Kosofsky Sedgwick argues: 'If such compulsory relationships as male friendship, mentorship, admiring identification, bureaucratic subordination, and heterosexual rivalry all involve forms of investment that force men into the arbitrarily mapped, self-contradictory, and anathema-riddled quicksands of the middle distance of homosocial desire, then it appears that men enter into adult masculine entitlement only through acceding to the permanent threat that the small space they have cleared for themselves on this terrain may always, just as arbitrarily and with as much justification, be foreclosed' (1990: 186). She goes on to suggest that 'the result of men's accession to this double bind' is the management of 'the threat' of homosexuality through both prescription and proscription (186).

'Take no prisoners' ethos aside, there have always been gay men who have trained at Le Studio, including those who were publicly and proudly out. Their incorporation into the group provides an opportunity to explore negotiations of diversification and fracturing in this community. Managing this difference requires the male corps to balance masculine solidarity ('Felipe' asserts: 'We are *all* "the guys."') with maintaining the corps' heterosexual, homosocial equilibrium, the 'middle distance' in this dancing community. At Le Studio, this balance is not accomplished through any overt or explicit pre- or proscription of personal identity or behavior. Quite the contrary. It is negotiated, instead, through the substitution of the danced body and its style of inhabiting ballet technique for any explicit acknowledgement of sexuality. One example from the mid-1990s is especially illustrative: 'New York School,' accompanied by extravagant, limp-wristed movements of the arms, was a devil term linking effeminacy in technique with the ridiculous geographic other. Two instances of this negotiation of difference in/and community are exemplary. Neither case involves simple pre- or proscription but rather a blend of both.

One gay long-standing member of the Le Studio community, 'Nicholas,' was folded into the 'golden age' by a consensual, accommodating fiction, one he both acknowledged and self-reflexively abetted. It was widely acknowledged that 'Nicholas' was 'different.' This difference was coded and publicly explained as 'older' and 'quiet.' I was interested in the formulation 'quiet,' and assumed it was meant as a euphemism for 'closeted.' When I asked, I was told by one male company member that it meant 'he's not a pig like the rest of us,' and by another that it meant 'he doesn't hang out with us.' In addition, 'Nicholas' was regarded as a 'hard worker' who was not 'a whiner.' In this he adhered to the company ethos of consistent, visible effort put into one's dancing.

At various points in my research during this period, three of the men approached me individually to reassure me that Nicholas was 'okay' and 'a good guy.' 'Nicholas' himself regarded this with a fair amount of amusement: 'They know. I know they know. They know I know they know. With all the knowing around here, it doesn't make a lot of sense to make statements, and I'm not interested in doing that anyway.'

'Jonah,' however, presented a more extreme challenge to the Le Studio men. He was 16 when he arrived at the school in the waning days of the 'golden age.' A slender, pale, patrician-looking young man, he had considerable previous training. He never self-identified as gay. What was more important than his real or perceived sexuality, for this community's purposes, was what they considered his 'effeminate' ballet style. The men of the company saw him as a challenge; in class, they were goal-directed about his socialization into the group. This involved altering his 'effeminate' technique and, in their words, 'breaking him' of what was widely perceived as his laziness and his narcissism. This latter was inferred by what the community saw as his excessive preoccupation with the mirror.

In class, the men were patient and supportive, as much to demonstrate their collective masculine *noblesse oblige* as out of genuine empathy for a dancer whom they saw as talented but unmotivated. He was seen as 'throwing it away,' with 'it' referring to both his seemingly natural abilities and the constructive criticism he was given to maximize them. Out of class, they grumbled about 'Jonah' as if they were frustrated by an odd and recalcitrant pupil. Some found it particularly galling that a number of the girls in the company had crushes on 'Jonah' and saw him as a kind of moody, Keatsian romantic figure. Among the men, he was compared unfavorably to 'Michael,' 'one of our own,' who, while not having 'Jonah's' facility in some aspects of technique, was seen as infinitely more worthy of esteem because of his work ethic. When 'Jonah' was corrected in class, he would often stop and sit out, pout or leave; 'Michael,' on the other hand, would persist, even after slips and falls:

> *JH*: Well, I've been watching ['Jonah']. He does some things well.
> *'Russ'*: Pose. He poses well. New York School.
> *JH*: Well, he has good extension and he can turn.
> *'Russ'*: Here's a guy who's making a decision, you know? 'Do I take this road or that road?' I've seen guys after they made the decision but this guy, he hasn't made it. You can see it. We're helping him out; got to help him out. He's young. If he makes it, he can really, you know, *make it*. But if he takes the easy road and just stays where he's at – I don't know. He won't get any better. He's got to push it now or never.

I didn't know what 'Russ' meant by 'making a decision' about which 'road' to take until his very last sentences. Was he talking about choosing one's sexuality or one's level of effort in ballet training? In some ways, the confusion was illustrative, as was the assertion that 'We're helping him out; got to help him out.'

The pedagogical frame surrounding and containing 'Jonah' was thick and pervasive. In the eyes of the company men, it needed to be. They had a considerable investment in how 'Jonah's' and, by extension, their own, homosociality was read from within and without. The most explicit example of discursive containment I had seen involved 'Felipe' explaining and apologizing for 'Jonah' to Charles Maple after a disastrous attempt to integrate him into a performance of *Fall of the Feet*, as noted above, the men's signature piece: 'He's an okay guy, really, buddy. But he doesn't always get it. Right? He doesn't always get it. We're working on him, me and 'Russ.' He's young, buddy. And he's okay, you know? He's lazy, a little lazy, you know, till [the *pas de deux* instructor] kicks his ass for, you know, not trying. He's used to resting on it. We're working with him, though. He's, you know, basically okay.'

Note the parallels between the terms used to characterize both 'Jonah's' and 'Nicholas's' 'difference.' One is 'young,' 'doesn't get it' and is in need of pedagogical intervention. One is 'older' so such intervention is both inappropriate and unnecessary because Nicholas 'works hard' and, perhaps, because he is 'quiet.' Both, though, were 'okay.' For the masculine homosocial contract to be in force, particularly among the men in the company, difference must be simultaneously contained and 'okay.'

Managing real or perceived differences in sexuality was not the only challenge facing men and boys at Le Studio. Ironically, another came from the very process that 'secures' homosociality from homosexuality: the traffic in women's bodies. In ballet, this traffic is literal; men lift and support women, assuring heteronormativity on stage by anchoring it in the romance of the couple. If, in such partnerings, the young women gain corporeal agency in spite of how the partnership appears on stage, as noted above, the men exude agency. They make things, lifts particularly, happen. Yet this is an ambivalent agency. 'Mark' observed: 'Well, they need us to hold 'em [the women] up. We get 'em up, right? Right. We get 'em up. Sometimes, though – not here so much, I think because of Phip and Chip – but sometimes at other places I dance at, I have to take a break because I – you – you start feeling like a tow truck, you know?' 'Kyle' added: 'That's why I *love* the men's piece [*Fall of the Feet*]. At South Bay [his former company], they were cool and I did some good stuff. But I felt – after awhile you feel like you're just a crane.'

'Mark's' and 'Kyle's' observations address an underlying ambivalence of masculine agency in ballet: to be an 'able-bodied male' (Hearn, 1992: 4–5) is to find pleasure in getting them/it up; function equals agency. And yet, over time, the burden of 'getting 'em up,' of servicing the young women whom they are supposed to dominate in the *mise en scène* of the choreography, coupled with the increasingly insistent view of self, not as agent but as vehicle, becomes disconcerting. The men come to feel, in 'Mark's' words, like 'just a big set of delts,' or, as 'Don' once observed walking into *pas de deux* class, 'Here I am: upper body strength.'

It is interesting to consider, in this light, words of the male ballet dancer's seeming *doppelgänger*, the 'Terminator,' Arnold Schwarzenegger: 'You don't really see a muscle as part of you, in a way. You see it as a thing. You look at it as a thing and you say well this thing has to be built a little longer... and you look at it and it doesn't even seem to belong to you. Like a sculpture' (in Goldberg, 1992: 176). As Jonathan Goldberg observes, this is the 'not belonging part, movable thing,' activated by an ambivalent agency in which one is both actor and acted upon. As 'acted upon,' the male dancer then occupies the structural position of the 'good girls who are easy to find,' as opposed to the actor who is 'the show.' If this 'movable thing' is read as a stand-in for the phallus, then here it is sutured to a male body poised between 'getting 'em up' like a man on the one hand, like a tow truck on the other. Here, the masculine agency that seemingly stabilizes heteronormativity in this dancing community both exceeds and falls short of guaranteeing the reassuring, coherent, authoritative subject position it appears to secure.

Like the girls who forge a home in ballet technique at Le Studio, a diverse community of men and boys use this place as 'an analeptic – a creative solution' that allows those who train there to 'forget our separateness and the world's indifference' (Tuan, 1992: 37, 44). In very specific ways, however, this is a fraught solution as much as a creative one. David Mamet suggests, in an essay entitled 'In the Company of Men,' that: 'the true nature of the world, as between men, is, I think, directed towards the outside world, directed to subdue, to understand, or to wonder, or to withstand together, the truth of the world' (1989: 90–1). For this multi-ethnic community of men, part of what must be subdued and managed is not any essential 'truth of the world,' but rather the internal tectonics of masculinity itself. The community of effort these men and boys construct subsumes ethnic, class and age differences into a consolidating fiction; they are *'all* the guys' who 'take no prisoners.' This fiction consolidates in part because it excludes: there are no women buddies.

Yet women's bodies both set masculine agency in motion and, by rendering this agency merely structural, simultaneously undermine it. As their agency is rendered ambivalent by the bodies of the interchangeable 'good girls' who are 'easy to find,' the male corps is thrown back on a fault-riddled homosociality, one that raises, in Sedgwick's words, 'the potential unbrokenness of a continuum between the homosocial and the homosexual' (1985: 1–2) in an activity already coded and read as feminine/effeminate. Even more theatrical spectacles of masculinity, an anxious disciplinary pedagogy, or both may be inserted to secure a workable 'middle place' on this continuum. The challenge of the latter is to 'socialize' differences like 'Jonah's' while maintaining that they are 'okay' because 'we are *all* the guys.' The challenge of the former is that such displays might yield 'reprisals' and 'debts to be paid,' further fracturing solidarity rather than shoring it up. Thus, male heteronormative subjectivity, forged through ballet as an art of existence at Le Studio, is always already making an unsettling spectacle of itself or, perhaps more accurately, it is a spectacle that unsettles itself around the bodies it shelters and supports.

Yet, for these men and boys, the sublime culmination of their efforts to 'subdue' ballet technique, or under- or withstand it, is just as freighted as it is for the young women who 'find and make something amazing' onstage, though they describe it in different terms. After watching 'Michael' take class at Le Studio for years, I saw him dance the role of the Nutcracker Prince with the company. As he emerged in character from the gaggle of little toy soldiers surrounding him, he seemed to shine. I was so moved by his visible, completely palpable joy that I burst into tears. When I mentioned this to him briefly backstage, he smiled, a bit embarrassed by my reaction, I suspect. His response, like the words of 'Joanna' and Erica, stayed with me well after he left to begin his professional career: 'I just felt I belonged up there. It all came together and I just felt strong – and really quiet – like I knew I was right where I wanted to be and that was it. I don't even remember parts of the show beyond just feeling like this is was like it should be.'

*   *   *

Whatever its strategic ambitions or popular profile, there is no one perfect place, time or body in ballet training. Despite its monarchical heritage and its use of French, regardless of its rigorous mapping of the body and its impatience with age, men and women, boys and girls insert themselves into it with thousands of idiosyncratic local turns in multiple vernacular landscapes. Sometimes they find injuries, anorexia and frustrated ambitions. Sometimes they roam for ballet. Sometimes they find a home, perhaps even a utopia, in it, if only for a moment when 'it all comes together.'

At Le Studio, ballet is an art and an urban art of existence. The technique purports to transcend difference, even gravity. The art of existence does not. Instead, in myriad small rituals of solidarity, in closeness and in competition, teamwork and talk, selves are created and exposed, sheltered and unsettled. In one sense, amateur and semi-professional ballet practice makes small things, including local performances and some professional careers. Yet it also 'changes what one even thinks is possible in the world,' not by reproducing dancing exotics in fantasy worlds but by offering opportunities – sometimes complicated ones, but opportunities nevertheless – to forge communities out of hard work and beauty, pleasure and pain, critical sorority and uneasy fraternity.

Soyini Madison asserts that 'all honest words' about creativity and its products are 'like rain on fertile ground' 2000: 315). 'Michael's,''Joanna's' and Erica's words about ballet are honest ones, which does not mean that their assertions of transformative moments onstage and off exist outside of all of the social, material complexities that make such moments possible. In such moments, dancers share the pleasures of remaking the global city, their bodies and, indeed, the world, as 'it all comes together' in performance. But, as Erica suggests, these 'gifts' are secondary. What keeps her coming back to the technique, and students, including 'lifers,' coming back to Le Studio, is something more profound. This is the actual, difficult, daily practice of remaking themselves and each other as bodies; as communities however riven by internal contradictions imperfectly managed; as memories; and finally, perhaps, as paradigms that suggest other possibilities of coming together in labor and in difference to find or make something amazing.

3
'Saving' Khmer Classical Dance in Long Beach[1]

[T]he Khmer Rouge surrounded Phnom Penh, and on April 17, 1975, after five years of civil war, they took control, waving their flag in the streets. Until January 1979 they forced all Cambodians to live in labor camps and work fourteen-to-eighteen hour days. They fed us one daily bowl of watery rice; they separated families; they destroyed all Cambodian institutions and culture; they systematically tortured and killed innocent people. It is estimated that during this time nearly a third of the Cambodian population was killed due to disease, starvation, or execution.

(Dith Pran, 'Compiler's Note.' *Children of Cambodia's Killing Fields: Memoirs by Survivors*, 1997)

[F]or those who undergo trauma, it is not only the moment of the event, but of the passing out of it that is traumatic; . . . survival itself, *in other words*, can be a crisis.

(Cathy Caruth, 'Trauma and Experience: An Introduction.' *In Trauma: Explorations in Memory*, 1995)

Not every dancing community is successful or productive. Not every dancer finds, or wants to find, a material community of fellow practitioners. Sometimes the challenges of life in the global city, particularly poverty and dislocation, preclude solidarity through art-making; sometimes personal damage and despair do. Sometimes, communities, families and individuals fracture under the weight of all of these and the cohesive power of dance technique is not strong enough to make repairs.

Dance technique offers communities protocols for reading the body; technologies for fashioning subjectivity and solidarity; and an archive, inserting all of these into an accessible, repeatable history. As archive, technique contains and organizes the traces and residues dance leaves behind, and out of which it forms again: injuries, vocabulary, relationships. Here, technique-as-archive becomes what Paul Connerton calls an 'inscribing practice.' It 'traps and holds information long after the human organism has stopped informing' its individual enactments (Connerton, 1989: 73). But this archive, and its residues and traces, its inscribing practices, are not only cognitive and corporeal, as Connerton might lead us to believe. It is not just physical labor that constructs specific archives of technique for dancers, not simply cognitive labor that fills them, but also the unrelenting dailyness of emotional and relational labor. And further, like many of the emotional and relational labors that constitute them, these archives are haunted affairs.

'Haunted places are the only ones people can live in,' writes Michel de Certeau (1984: 108); he might have observed, as well, that they are often, for good and ill, the only places that people can perform in. What follows is an account of one very small archival hauntopia. Certainly it is not representative in any simple way. Yet it has persisted in my mind and heart long after (perhaps because) my contact with the individuals involved had ended.

The taxing, relentless, affective labors of technique often lend themselves to hagiography, to hero stories of what performance overcomes and to the triumphs of dancing communities constituted in the face of adversity. This account is not one of these. Rather, I suggest here that, while performance often overcomes, it is just as likely to succumb to exhaustion, to deep personal anguish and profound social dislocation, and to the memories it has deployed as both blessing and goad.

Answerable bodies/unclaimed experience

I begin by acknowledging, explicitly, the fraught relations between history, trauma and truth that undergird the following discussion. While these relations are the living and breathing of the psychoanalytic enterprise, they are much less readily acknowledged in an ethnographic one, and particularly one so intimately linked to genocide, to its imperative to witness, to speak the truth back to power, as in recent memoirs by the survivors of the Khmer Rouge (see Criddle, 1987; Him, 2000; Nath, 1998; Ngor, 1987; Pran, 1997; Szymusiak, 1999; Ung, 2000). Perhaps scholars of performance are ideally positioned to engage such fraught relations, to 'constellate . . . various approaches to the nexus of performance [trauma, truth and] history' (Pollock, 1998: 1) while mindful of the promises of, and lacunae within, representation in its inexorable trajectory toward disappearance. But the relations I mean challenge even the capacitating frame of performance. To paraphrase both Emerson and Geoffrey Hartman, I suspect my instruments (Hartman, 1993: 243), both theoretical and interpersonal. Simply put, the crises of truth so central to the relationship between trauma and representation undermine, at almost every turn, my attempts to narrate a coherent account of one family of Khmer survivors, of how they told themselves that dance was the reason for their survival, and of the ways they tried and failed to find solace therein to performatively shore up survivorship and continuity. These crises of truth in trauma remind me that cultural gaps, as well as representational and methodological ones, are fundamentally human – that is to say, affective, psychic – gaps, and they divide husband from wife, parents from children, and family from community, as well as researcher from the ease and stability of scholarly claims.

Thresholds of secrecy

I met the Sem family through their son at a performance of the Classical Dance Company of Cambodia in Los Angeles in 1990. The children were fairly gregarious and suggested that, as I was interested in classical dance, I should meet their parents who, in their words, 'danced *real* Cambodian dance in Cambodia and in [refugee] camp.'

'Sem' is a pseudonym and my decision to use it requires both an explicit acknowledgement of, and my complicity in, the 'thresholds of secrecy' (Feldman, 1991: 11) which characterized my relationship with this family. I use the pseudonym, in part, because, as I will explain later, I am not able to ask the family directly for permission to represent them and their stories, something I generally do with dancers I write about. Beyond this, I have been unable to account for and confirm key details about the family, which has led me to suspect that the name they lived under in Long Beach, California[2] was itself an enabling fiction, its utility ensured by the lack of subtlety for Khmer inflection in Western ears, and by wide variations in nonacademic transcriptions of Khmer into English.[3]

I have other reasons to suspect that the name under which I knew the family was what Allen Feldman might call a tactical 'spacing between discourse[s], between words and acts,' or past and present, one of many recognitions of the impossibility of psychological and historical parity between myself and my interlocutors (1991: 12). Mail, some of which I was asked to decipher, came to the family's apartment addressed to different names. A search of licensing records and city permits for the donut shop in which, I was told, the Sems were partners turned up no one by their name, though it is perfectly possible that their use of 'partnership' was not a legal one. Father Ben Sem and his wife May were vague about the date of their arrival in Long Beach (1981 or 1982) and would not reveal who had sponsored them, though they regularly received generic mailings from a Protestant denomination active in refugee resettlement. Searches of relevant archives and records did not clarify matters. Ultimately, for reasons which, I hope, will become clear, I abandoned what I had always rather shamefully considered a 'detective operation' to ferret out some simple 'truth.'

During the two years I knew them, I was never fully clear about how they perceived me or my interest in Cambodian classical dance, or in their lives. At times, Ben in particular seemed to resent my visits; at others he seemed grateful for the attention. May and her daughters seemed to enjoy my company, though it was never clear if this was due to me personally, or to their relative isolation. Though I often brought groceries, or offered rides, these gestures did not seem to correlate, in any simple way, to my reception. Whether they were seem as some sort of *quid pro quo* by the Sems is impossible to say.

I visited the family at least once a month, generally every two weeks, staying for between one and three hours. Though I stated my interest in learning about, and writing about, Khmer classical dance, I am not sure they fully understood what that meant. The gaps and ambiguities that characterized our relationship from beginning to end could have been simply linguistic; they also could have been colored by family secrets too dark to tell, especially to a relative stranger, or to deep damage, borne of trauma, that bled over into every interpersonal, familial and communal exchange. Probably it was some combination of all of these.

Ben and May had met, they said, in 'Camp 2,' which I presumed at the time was Site 2, the largest of the camps along the Thai border.[4] The name used by the family while in Long Beach is a common one; it is represented in the Cambodian Genocide Archives and databases among both victims of the Khmer Rouge and the Khmer Rouge themselves. This is true of the pseudonym I have chosen as well. I have used pseudonyms to represent their daughters' 'American names,' Sandy and Jennie, which, at the time I knew them, they much preferred, to the pain of their mother who was given the names of all three children, in strict order, by a seer in the camp. This choice also reflects the distance between parents and children, a distance that figures prominently in Ben's and May's deployment of dance to theorize their own survival and legacy. I have used a shortened version of a Khmer name, 'Rith,' similar to that of the Sems's son, their oldest child, in keeping with what was, at the time, his preference.

As Feldman reminds us, it is a mistake to view thresholds of silence, like the Sems's tactical ambiguities, as purely individual matters, though they are that. Like virtually every other first wave Cambodian refugee family, the Sems did not escape the Khmer Rouge autohomeogenocide intact or unscathed, physically or psychically. Like most, the Sems's survival, and indeed, their sanity, depended upon relational amnesia and aphasia. Ben Sem observed: 'You had to forget yourself, understand? Forget you were student, forget you have family, tell them [Khmer Rouge] you are farmer, you are builder. Then more forget. Forget hungry, forget tired, forget scared' (See also Ngor, 1987: 239–43; Szymusiak, 1999: 20; Ung, 2000: 32). In a very real way, the Khmer Rouge enforced a tyranny of forgetting, from the macro to the micro: the calendar began with Year Zero, monuments and archives of all sorts were systematically destroyed, families split apart. Ben Kiernan notes, 'The Khmer Rouge hoped to use children as a basis of a new society without memory' (in Pran, 1997: xvii).

The Sems's 'inability to remember,' and/or their unwillingness to divulge, key aspects of their autobiographies reflect a logic of memory in the 'death world,' a logic articulated by Edith Wyschogrod and Elaine Scarry. Wyschogrod uses the term 'death world' to characterize a form of social existence, 'in which vast populations are subjected to conditions of life simulating imagined conditions of death, conferring upon their inhabitants the status of the living dead' (1985: 15). Wyschogrod describes this phenomenon as unique to the twentieth century, and in particular, to the Holocaust, because of greater technological mediation. Her account of the death world resonates so clearly with those of Khmer survivors, however, that I find her characterizations useful, despite the comparatively low-technology genocidal practices of the Khmer Rouge. Intrinsic to both contexts is the image of the inhabitants of the death world reduced to animated corpses, bodies without sentience. This radical reduction to the brute circumstances of survival, and particularly the emphasis on all-consuming hunger, runs through all published accounts of Khmer survivors.

As Foucault tells us, disciplinary, carceral systems work by etching their imperatives onto the body and so it is, *in extremis*, in the death world. The dynamics here are two-fold. First, the life world, in material and discursive terms, is radically compressed, as Wyschogrod describes: 'It is the experience of the boundary, of crossing over from free space to forced enclosure. The sealed off transports with their human cargo, the vast hordes . . . supervised by armed guards, constitute the first restriction of mobility. Intolerable levels of population density continue and are systematically enforced by the . . . system itself. The crowding which characterizes conditions of eating and sleeping is attenuated only by regimens of meaningless work' (1985: 19). While transports were not used to enable the autohomeogenocidal work of the Khmer Rouge, survivors' accounts, and particularly those of individuals and families evacuated from Phnom Penh, emphasize these elements of crowding, of armed soldiers enforcing seemingly irrational movements of large populations, and, particularly, back-breaking labor coupled with vicious hunger. As Wyschogrod concludes, 'At the levels of vital existence the life-world suffer[s] a severe condensation of meaning' (1985: 20).

This condensation of meaning is etched onto the body through pain, through beatings, rapes, torture that are the physical signifiers of the authorities in the death world. I use 'signifiers' ironically for, as Elaine Scarry demonstrates, torture/pain undoes the very process of signification: 'Physical pain does not simply resist language, but actively destroys it, bringing about an immediate reversion to a state anterior to language, to the sounds and cries a human being makes before language is heard' (1985: 4). Further:

> [t]hough the capacity to experience physical pain is as primal a fact about the human being as is the capacity to hear, to touch, to desire . . . it differs from these events . . . by not having an object in the external world. Hearing and touch are of objects outside the boundaries of the body as desire is desire of x, fear is fear of y, hunger is hunger for z; but pain is not 'of' or 'for' anything – it is itself alone. This objectlessness, the complete absence of referential content, almost prevents it from being rendered in language.
>
> (Scarry, 1985: 162)

Nadezda Mandelstam observed, 'If nothing else is left, one must scream. Silence is the real crime against humanity' (in Wyschogrod, 1985: 21), and herein lies a minor triumph of the death world. Voice, language, even screams, may recuperate shards of the lifeworld but, as you will hear from Ben and May Sem, pain precludes language and appropriates the scream as an object unto itself, indeed, particularly for the torturers, one that legitimates itself.

Simply put, in the death world, the discursive apparatus of memory is as likely to fall into a pit of irretrievable inwardness, inaccessible, as it is to generate an imperative to 'tell.' To say that, once freed from these immediate exigencies, the Sems's reluctance to present coherent auto-biographies might be a simple, tactical performance of mistrust is both true and misses the point. Such a reading ignores the social surround that regulates and polices personal memory, which, indeed, makes such memory possible. Ben and May Sem *had* to tactically manage memory because, to paraphrase the words of Adrienne Rich, they always ran the risk of remembering their names and, in so doing, failing the strategic forgetfulness which kept them alive. In genocide's aftermath, memory does not always equal information retrieval or expressive force, much less solace. It was rather, a wound and, moreover, one whose unique etiology was inextricable from deeply personal limits and frailties.

Once in this country, strategic amnesia proved difficult to abandon. Some stories could never be told, as Sandy Sem illustrated: 'We had some project for school, to talk about our culture and our families – like grandmothers and grandfathers and stuff. But you can't ask them [her parents] about that, him [her father] especially because he gets mad and the teacher – right – she's going to believe that. And I'm going to go in, okay, and say: "My family's from Cambodia and everybody's dead from the war or over here somewhere, but nobody says, okay?" So I just made it up.'

In a macabre way, this strategic amnesia is reflected in the Khmer Rouge's genocidal logistics. Among the thousands of photographs of victims executed in the infamous Tuol Sleng prison, themselves repres-enting only a fraction of the regime's casualties, there is no correl-ation between the individual photos, names and lists of confessions (Cambodian Genocide Project). The imprecision of this documentation stands in direct proportion to its quantity; Seth Mydans observes, '7000 or more portraits [were] intended as a bureaucratic record of prisoners entering Tuol Sleng. All the subjects died; only the portraits survived' (Mydans, 1999). Survivors' accounts mention files, dossiers and note-books to record 'confessions' or the testimony of informers. To possess a camera, or a notebook, or a book of any kind, however, could be a mark of an educated past and thus a death sentence (Mydans, 1999). At the opposite pole of power under the Khmer Rouge, artist Vann Nath, one of only a handful of survivors of the infamous Tuol Sleng prison, was forced to join a cadre of painters and sculptors whose task was to memorialize Pol Pot on canvas and in stone (Nath, 1998: 58–82).

Perhaps this bipolar response to documentary technology by the Khmer Rouge explains why Ben Sem became agitated, almost to the point of violence, when I appeared for our first meeting with a tape recorder. The intensity of his reaction frightened me so thoroughly that I only contacted the family again to make sure no harm had come to anyone as a result of my naive intrusiveness. I was surprised to be invited back. Thereafter, I took notes very discretely, and generally when Ben was not present. There were rare occasions, though, when Ben or May might say, firmly, 'You write this down. You write what I say.' I always did, and I did so when both, at different times, offered apologies of sorts, and explanations, for Ben's outburst over the tape recorder. May explained: 'He scare you – yes, I know. You write this now, so you learn. They [the Khmer Rouge] try to trick you – use your words. They go around and they write, they say: "Your auntie goes with Lon Nol soldiers, it say right here." I think: "dirty people [they] can't write or read" so I think there's no writing in those [mimes a pad] but people die from this. Yes. From the words – so you learn. No talk, no write don't have them use. He [Ben] think about that all the time. All the time.'[5] Ben later added: 'Khmer say, "Your tongue will steal from you." ' While I could never find this saying in any written list of Khmer proverbs translated into English, I did come across a number which expressed similar sentiments: indiscreet or imprudent use of language could lead to ruin (see, for example, Fisher-Nguyen in Ebihara, 1994: 102–3). As Ngor, Him, Ung, and other survivors have indicated, this was true, in the most brutally literal sense, under the Khmer Rouge (see also Kiernan, 1996: 170). Ben continued:

> *Ben*: We don't talk [about] that time. Far away now. Then too. Far away then. Like not [points to himself], not us, understand? Here, not here. Not live, just move. Not see – like – say?
> *Rith Sem*: Ghost?
> *Ben*: Yes.
> *Rith*: They were like ghosts.

In *Ghostly Matters: Haunting and the Sociological Imagination*, Avery Gordon writes: 'The ghost always registers the actual "degraded present" . . . in which we are inextricably and historically entangled *and* the longing for the arrival of a future, entangled certainly, but ripe in the plenitude of nonsacrificial freedoms . . . The ghost registers *and* it incites' (1997: 207). Accounts of Khmer survivors are replete with this paradoxical ghostliness so characteristic of the death world: reduced to shades themselves, they are both haunted registers of atrocity and avenging angels, impelled by other ghosts, even in diaspora, to reimagine the generative possibilities of Khmer culture.

For May Sem, this dual status – both ghost and haunted – was literal, as she confided to me after a year and a half of my visits. This was a long confidence, and a rare one, and I quote it at length. I couldn't determine, and didn't ask, if she was referring to her life under the Khmer Rouge, in camp, or to some blurred horror of both.

> *May*: You write this – they kill us even when we live. Even when we live we are dead. Dead bodies, working, all quiet. That's all. It's hard to say, but no one talked. What to talk about? Why this one [killed/died], this one, this one and not you? Talk about who sees the worst? Whose more family die? I tell you – no one talked. Me? I don't think. Don't think. Work, eat, sleep – that's all. In camp there was a man always talking, 'Wife die, mother die, house burn,' on and on, like that. On and on, everyday. We say, 'Shut up. Shut up with this talk.' But he don't and one day they kill him – kill him, I think, yes, because some men beat him up and then gone. No more. Why? Can't talk, can't hear. I dance. I'm not like my husband, good student like him, but I dance in camp. Music is beautiful, steps are hard and they are beautiful. I dance to keep music in my [ears] and keep talk out. Yes. And so there is some reason for me to be left – to help my husband so Khmer dance don't die too, like everything else beautiful in our country die. I hear them sometime.
>
> *JH*: Hear who?
>
> *May*: I hear my teachers who did not get out. I hear them sometime in the day. I don't see, just hear. I am scared but they tell me the step [movements generally] to the dance. I am like child, a baby. I listen to them tell me the step. At first I am so scared, I don't listen. Maybe go crazy like my neighbor I tell you. But I listen to them tell me the step, then I do. I do. Sometimes I do better, I think so, because they tell me. Maybe a little crazy, understand? But I do the step better.

As surely as memory shrouded some texts in silence, it resurrected others and, in the Sems's case, these other texts were dances specifically and, more generally, classical dance as an ur-text of Khmer culture.

The answerability of memory

May's ghostly interlocutors recall James Brandon's observation that, while 'performing arts do not die, performers do' (in Sam, 1994: 47).

Beginning in 1975, Pol Pot and his Khmer Rouge forces set out to system-atically and ruthlessly destroy classical Cambodian performance and culture. Benedict Anderson may be correct when he notes that classical Angkor, 'emblazoned on the flag of Marxist Democratic Kampuchea,' was a 'rebus of power' (1991: 160–1), but neither power nor piety spared the material bodies of the Apsaras, the dancers who served as incarna-tions of Angkor's reliefs. Pich Tum Kravel, and former Minister of Culture Chen Phon, estimate that 90 percent of Cambodian dance teachers and performers were murdered or starved to death; the exact number may never be known (Chen, 1993). Those few who managed to escape either remained in the country posing as peasants or, more commonly, made their way to refugee camps along the Thai-Cambodian border.

> *May*: They killed everybody. You can't even think so many. You old – killed. You student – killed. You dancer, play music – killed. We [referent unclear] are [presumably relatively] strong, we run and walk to camp. Sometime we see other student in camp but [holds up two fingers] teachers. Why? Old, dance long time, look like dancers – killed. And I hear in camp, you know, so many did not get out. Dancers and teachers from the school. I see and I hear. They did not get out. Dead. All dead. Before I get to camp I – What if I am only left?

It is this possibility, that of being the 'only left,' that impelled May to practice her skills and try to pass them on to her children, even as she shyly observed, 'I am not good dancer like my husband. He went to school long time, and is good dancer. I am just dancer. Just okay.' It is this same possibility that channels survival, and survivors' guilt, into a potentially transformative, yet deeply ambivalent, agent and agency (Myerhoff, 1992: 236). Ben Sem clung to this fraught agency with special, and as I will discuss below, especially problematic, tenacity, despite the exigencies of displacement in the United States. He explained: 'I was not the best student, not the best. But I am here, so I must do this work. Hard to dance here. Kids say, "Oh my friends go here. I go with them." Say, "Oh, TV's on now. TV." I say, "How you think?! Think like this?!" If we don't think like this [points to boom box playing Khmer music], if we don't think like this in camp there is no dance! No more!'

Barbara Myerhoff observed: ' "It can't happen to me," comforts on-lookers but not survivors themselves. They know by what slender threads their lives are distinguished from those who died; they do not see in themselves soothing virtues or special merits that make their survival inevitable or right. They know how easily it could have happened to them; to these people, complacency is lost forever' (1979: 24). And so Ben and May, 'not the best' student and 'just okay' dancer, deployed dance as a fictive ground on which they legitimized their survival and organized a sense of agency despite survivor and refugee status.

In *Spectacular Suffering: Theatre, Fascism, and the Holocaust*, Vivian Patraka engages Elin Diamond's notion of performativity and reimagines it in light of the Holocaust:

> According to this model of the performative, 'the thing done' is a kind of yardstick, a system of beliefs and presuppositions that has taken on an authority and become a hegemonic means of understanding. The 'thing done,' then, represents particular discursive categories, conventions, genres and practices that frame our interpretations, even as we try to perceive the present moment of doing. As we are in the doing, then, there is the pressure of the thing done. The doing is not knowable without the thing done, and the thing done is all the discursive conventions that allow us to think through a doing.   (1999: 6)

In the case of 'the Holocaust performative,' Patraka identifies 'the thing done as *the thing gone* ... It is the goneness of the Holocaust that produces the simultaneous profusion of discourses and understandings ... [T]he absoluteness of the thing done [gone] weighs heavily on any doing in the Holocaust performative. . . . [It] acknowledges that there is nothing to say to goneness and yet we continue to try and mark it, say it, identify it, memorialize the loss over and over. This is the doing to which Diamond refers, the constant iteration against the pressure of palpable loss . . .' (1999: 7).

While I do not want to understate in any way the specificity of Patraka's theorizing the Holocaust performative (per her observations on p. 3), it is striking to note how forcefully it resonates with the testimony of Khmer Rouge survivors, and with the Sems's commitment to dance, over against the goneness of classical Khmer culture in the aftermath of the Pol Pot regime. Yet, while the invocation of goneness is explicit in both general and highly personal 'introductions of [Khmer Rouge] atrocity into representation' (Patraka, 1999: 7; see especially Sophiline Cheam Shapiro's 'Songs My Enemies Taught Me' in Pran, 1997: 1–5), for the Sems, this animating sense of 'goneness' was expressed in general, cultural terms ('so Khmer dance don't die too, like everything else beautiful in our country die'), and only rarely and obliquely in personal ones ('We don't talk that time.'). For the Sems, what personal silence foreclosed, dance enabled; through the solace of good form, it offered an answer back, albeit a problematic one, to their pervasive sense of goneness.

Patraka writes that doing/performance is always accountable to the thing gone (1999: 7). It is also useful to imagine this accountability in Bakhtinian terms. Here, in the case of the Sems, dance becomes the vehicle through which Ben and May perform *answers*, both for their individual survival, and that of Khmer culture. At its simplest, Bakhtin's 'answerability' is very like accountability, predicated on an inter-animation of art and guilt, life and blame. In 'Art and Answerability,' his earliest known publication, he writes: 'I have to answer with my own life for what I have experienced and understood in art, so that everything I have experienced and understood would not remain ineffectual in my life. But answerability entails guilt, or liability to blame. It is not only mutual answerability that art and life must assume, but also mutual ability to blame... Art and life are not one, but they must become united in the unity of my answerability' (1990: 1–2).

Answerability is ambivalent. At its most generative, it seems to offer an ethical opportunity to deploy art to speak back, both to life and to goneness, as well as the reverse. At its most impotent, answerability seems to circumscribe agency, limiting it to reactions of singular subjects condemned to answer to, or 'rent' meaning (Hitchcock, 1993: 10). Central to this ambivalence is Bakhtin's emphasis on the mutuality of art and life congealed into a unity through answerability-in-practice. In the case of the Sem family, can answerability, as they deploy it ('I am here so I must do this work'), congeal into such a unity? Can art, life and goneness answer for or to each other? I suspect Ben and May, Ben in particular, would have it so. Such a unitary answerability animates many Khmer survivors, both in the United States and in Cambodia (see Turnbull for one recent example).

Yet it seems to me that this unitary emphasis in answerability, both in Bakhtin's terms and in Ben Sem's, can be held to a different light which reveals, instead, a palimpsest in which speech and silence, the dancing Apsara body and the precariously poised refugee-survivor body, etch themselves over and over onto each other, again and again, always incompletely, always in/as difference. I attribute this profoundly ambivalent palimpsest of answerability through performance in the Sem family to a peculiar relationship between a classical form and a fraught subject. Here, as I will explain, the dancer was always already too distinguishable from the dance, turning precisely on the pivot of guilt and blame so intimately linked, both to his own legitimation of his survival, and to Bakhtin's theoretical tool to describe it. Perhaps the most enabling way to explore this relationship is to take a cue from a character in Lucia Murat's film *How Nice to See You Alive*, itself concerned with surviving a murderous, authoritarian regime: 'the question [of survival and answerability] still goes unanswered. Maybe what I can't admit is that it all begins here – with the lack of an answer. I think I should change the question. Instead of "Why do we survive [and answer for it]?" it should be "How do we survive [and answer for it]?"' (Murat in Sippl, 1994b). That is, we can turn to specific technologies of answerability.

Technologies of answerability

Pure products

Joseph Brodsky observes: 'What the past and the future have in common is our imagination, which conjures them. And our imagination is rooted in our eschatological dread: the dread of thinking that we are without precedence or consequence. The stronger that dread, the more detailed our notion of antiquity or of utopia. Sometimes . . . they overlap . . .' (1994: 40). This confluence of antiquity and utopia is particularly magnified for the refugee or the exile, who faces 'two types of immanent and imminent threats simultaneously: the threat of the disappearance of the homeland and the threat of themselves disappearing in the host society. Fetishization [as a strategy for negotiating these threats] . . . entails condensing all the meanings of home . . . into substitute fetishes and frozen stereotypes' (Naficy, 1993: 129). Hamid Naficy concludes, the net result of this fetishization of past and future is that '[t]hrough controlling "there" and "then," the exile can control "here" and "now"' (1993: 132).

The fetish of Cambodia as it was and never was has an especially acute pull for refugee survivor artists, for whom answerability to and through Khmer culture is seen as rooted in a sacred, utopian past, mired in a degenerate present. No matter that the exact nature of this past is unclear as Khmer traditions are, to a great degree, contested constructs among Khmer themselves (see, for example, Ebihara, 1994: 5–9). Juxtaposition between sacred past and degenerate present permeates Khmer discourse, as does the sense that the future may hold only greater remove from utopian antiquity and, ultimately, perhaps abyss. This has been modified considerably by the death of Pol Pot and the collapse of the remaining Khmer Rouge forces, as well as a concomitant investment of both Cambodian-American dollars and commitment (see Pape, 2001;

and 'Cambodia Sentences', 2001). Still, in the early 1990s when I met the Sems, the rhetoric of these juxtapositions between present and past was both radically essentialist and pessimistic, as these examples illustrate:

> *Chen Phon*: Khmer culture was the most refined, the most gentle of Indochina. The most beautiful. It lasted 1000 years. Now Khmer can't add, just subtract. Can't multiply, just divide. In Cambodia now there can only be subtracting and dividing.
>
> (Chen, 1993)

> *Ben*: I tell my children, 'Khmer culture so beautiful. Artist and dancer live well because they are like jewels. Live like jewels.' My teacher say, 'Every dancer a book. Every great dancer a . . . a . . . many books. Every great dancer a . . .' Say?
> *Rith*: A library.
> *Ben*: Every great dancer is a library. Khmer Rouge kill the dancers, kill the books. All the books are killed.

Indeed, some Khmer viewed the period of 1975–79 as the beginning of the end of the Buddhist era, the beginning of the extinction of all that is familiar as Khmer as part of a karmic cycle (Ebihara, 1994: 3; Smith, 1989).

If answerability, predicated on the reconstitution of Khmer culture, seemed logistically problematic if not theologically doomed, this did not keep Ben and May from trying, or from modeling their efforts on a 'pure product,' a cultural fetish, even if their approach to it was necessarily asymptotic. Literature, lectures and performances authored by Khmer artists are replete with assertions of this necessity for reproducing the pure Khmer product/fetish, even as they freely admit there are few existing models to which reproductions can be compared, and even as they acknowledge the exigencies which disrupt such reproduction. Consider ethnomusicologist and musician Sam-Ang Sam's discussion of the contingencies surrounding the assembly of a *pin peat* musical ensemble:

> In performances, Khmer musicians (myself included) both consciously and unconsciously borrow instruments from other Khmer ensembles and mix them with those of *pin peat*. Because of the scarcity and distribution of musicians and musical instruments this borrowing is difficult to avoid. In addition, Khmer traditionally hold great respect for elders. In performance situations, we cannot

tell our older musicians... not to play with us just because they do not know how to play a *pin peat* instrument. Moreover, excluding them from the group would reduce the ensemble even further, thus making it impossible to produce a full accompaniment. Worst of all, discouraging other musicians from playing might be seen as a break in the continuity of Khmer musical life.   (1994: 45)

Yet 'fidelity' was still the god-term for Khmer artists generally and, for the Sems, the only worthy answer to 'how do we survive and answer for it?' Variation and experimentation, even by artists with impressive classical credentials, is met with uncertainty and ambivalence at best (see Mydans, 2000).

In Bakhtin's view, the ethical, answerable performance is effective, regardless of contextual contingencies. Indeed answerability rehabilit-ates such contingencies and reinvigorates them, generating a coherent, unitary assemblage that is concrete and historical, affective and ideolo-gical. In *Toward a Philosophy of the Act*, which I see as a companion to, and elaboration on, 'Art and Answerability,' Bakhtin writes:

> This answerability of the performed act is the taking-into account in it of all factors... as well as of its factual performance in all its concrete history and individuality. The answerability of the actually performed act knows a unitary place, a unitary context in which this taking-into-account is possible – in which its theoretical validity, its historical factuality, and its emotional-volitional tone figure as moments in a single decision or resolution. All these moments, moreover..., are not impoverished, but are taken in their fullness, and in all their truth. The performed act has, therefore, a single plane and a single principle that encompasses all those moments within its answerability.   (1993: 28)

That is, within the architectonics of answerability, individuals, from our own unique spatio-temporal perspectives, perform, consummate, and, in so doing, unite moments of content, context, affect, and so on, in all their truths, as acts of authorship for which we are then answerable and for which we have no alibi. From this perspective, the relationship between authorship, answerability and performance in the Sems's deployment of the Khmer cultural fetish is especially complex and peculiar.

As Elin Diamond observes, incommensurability lies at the heart of mimesis: '[t]he body is never fully subsumed in impersonation,'

(1997: 180) or in mirrors or stories, however compelling. And between the utopian antiquity of Khmer dance, and the trauma and goneness that the Sems's deployed dance to answer, this incommensurability was especially acute. In contrast to Bakhtin's 'single plane, a single principle,' in contrast to an answerable unity between art, life and goneness, the Sems depended on an answerability which reiterated the deferral of life to the transcendent rhetoric of the dance. Mimesis became too full and too empty at the same time.

Virtual bodies

For this incommensurable element of answerability to become clear, it is useful to look even more closely at the specific virtual body organized by technique and ventriloquated through the material body of the dancer, that is, at the referent of this mimesis. Khmer dances function as narrative technology to embody and reproduce both sacred and secular folk texts and images, including the Cambodian 'Ramayana' and Buddhist and Khmer myths.

> *Ben*: We are book. Not just pictures like [holds up a magazine]. Not to look at only. We are – dance story, stories, not only nice good pictures.
> *Jennie*: It's like stories in a book. Some are funny and some are scary – not to me they're not but I guess to somebody, maybe over there [Cambodia] a long time ago, and some are really long.

Over the last 70 years, the Khmer classical repertory has been distilled; new dances have been 'extracted' from the larger epics and stand on their own, much as a solo or *pas de deux* is often presented independently of the ballet from which it was taken. The rituals and dance dramas, many of which traditionally span four to six hours or more, have been shortened. These compressed performances present an epic's dramatic highlights – a pivotal seduction or wedding, perhaps, or battles between goddesses and ogres. Even when illustrating moments of intense action, a battle for example, the effect of the dance is languid, more a tableaux than what ballet would present as narrative momentum.

Ideally, the visible body of the classical dancer, as author of the pure Khmer product, attempts to write itself over an invisible personal text ('we don't talk that time'). At the core of this act of authorship is mimesis of the most rigorous kind. Sandy, the Sems's oldest daughter explains: 'There is only one right way, only one right way. You learn like in a mirror, repeat it like the teacher is your mirror, okay? It's hard cause there's just one right way, like math.' This is not an idiosyncratic view. Cambodian dance functions linguistically, with an extensive vocabulary and syntax.

To reproduce classical stories, the body of the dancer becomes a complex intertext as the eyes, head, fingers, wrists, torso, legs and feet all speak through their specific, fixed repertory of poses at the same time. These positions and poses are highly stylized, demanding a level of precision and artifice that makes ballet seem naturalistic by comparison. For example, fingers and wrists are generally both fully flexed *and* fluid, with the former curving back away from the palms in a manner similar to, but seemingly even more stylized and extreme than Indian mudras. To achieve the finger curvature required, girls must continually bend their fingers backward from an early age. May continually urged her daughters to keep working for 'beautiful hand.' Likewise, movements for the feet often require toes to be flexed upward and held for the duration of the pose or sequence of movements. Jennie, in particular, would often parody this position, clomping about the apartment with toes upturned when she was prodded to practice, in sharp contrast to the silent, liquid changes of weight required of Apsaras.

The task of embodying the Khmer ideal is inflected differently based on gender. Further, the virtual bodies of the repertory are gendered, both essentially and performatively. According to Pich Tum Kravel, former director of the Department of Dance, University of Fine Arts, Phnom Penh: 'Women traditionally play all the roles in Khmer classical dance – drama, even male characters and ogres. Only in this [twentieth] century were male dancers allowed by the dance's royal sponsors to perform along with the women... To this day, though, male characters and demons are played by female dancers' (1990: 3). Women performers embody 'Apsaras,' celestial dancers who guard the heavens, serve as intermediaries between the sacred and the secular through their dances, and epitomize the beauty of Khmer culture. Indeed, Apsaras are at the very center of the Khmer cultural imaginary, immortalized in stone reliefs at Angkor Wat, the twelfth-century temple complex at the heart of the classical Khmer empire. Legend has it that Apsaras were born when gods and demons, locked in battle, churned up the foam in a sea of milk. The serene, opaque facial expressions of the dancers function to recall these divine beginnings and the Apsaras' celestial position. Male dancers, and women as well, perform acrobatic 'monkey roles' in sacred epics, or ogres and trolls in folk pieces called *Lakhon Khol*. Their movements are both more forceful and somewhat less refined than those of the Apsara role.

May: Girls have to start from very small because to train the body, you . . .

Sandy: You have to bend your hands [fingers] back all the time like this and be able to move them [fingers] separate[ly] or you can't do it right, and you have to know how your feet and eyes go.

May: Boys do later, start later. They do monkey dances, not so fine dances, like that.

Sandy: I think they [boys] have it pretty easy because they just run around and jump around. I guess not really but that's what it looks like. And girls have to [mimics mincing head, eye, hand movements].

In addition to mastering the vocabularies for head, eyes, arms, fingers, feet and stance, Khmer dancers must master the texts, music and movements of an extensive classical repertory. Because, as noted above, women frequently perform men's roles as well, they must master these intricate dances from multiple perspectives.

While the monkeys and ogres have their own transcendent aspects, the Apsara is the ultimate virtual, corporeal ideal and so merits special theoretical attention. In representing and ventriloquating the Apsara ideal, the body of the female dancer becomes the narrative container of the cosmos, a microcosm of myriad, nuanced interrelationships between the physical and the spiritual. This microcosmic body, in turn, has consequences for the visible, answerable body of the dancer. Susan Stewart observes that, in order for the body to 'stand in' for the universe, 'it must itself be exaggerated into an abstraction of the ideal. The *model* is not the realization of a variety of differences. As the word implies, it is an abstraction or image and not a presentation of any lived possibility' (1993: 135; emphasis in original). In Bakhtinian terms, this is not a novelistic body, but an epic one: 'Whatever its origins, the epic . . . is an absolutely completed and finished generic form, whose constitutive feature is the transferal of the world it describes to an absolute past . . . The epic past is called the absolute past for good reason: it is both monochromic and valorized (hierarchical); it lacks any relativity . . . that might connect it to the present' (1981: 15).

Bakhtin asserts that the epic is walled off from the present of its singer. Khmer dancers assert, however, that they *become* Apsaras. The difference may be that the Apsara is envisaged, by the absolute, hierarchal past, as always already an embodiment, rather than as a text.

As a technology predicated on and maintained by the epic body, the Apsara 'implicitly denies the possibility of death – it attempts to present a realm of transcendence and immortality' (Stewart, 1993: 133). And herein lies an additional ambivalence embedded in the Apsara ideal and classical dance as an answerable act of Khmer survivors. The Apsara, as a technology of answerability, is set in motion by the materiality of that which it implicitly defers, if not denies: death and displacement in the degraded present. Further, the Sems, the individual bodies who employ this technology, defer this too ('We don't talk that time. Far away.'). Yet both death and displacement do make their contestatory presences felt in 'What if I am only left?' I, as an outsider, was left with a palimpsest of answerability in which the body of the dancer writes over the invisible text of the refugee and, in so doing, is written over by the ideal body of the Apsara: a dance in which each move/body was a partial erasure and deferral of the previous one. The net result is not unity but the mutual destabilization of art, life and goneness, an exposure of their incommensurability, a destabilization and exposure the Sems sought to shore up through the dance.

Incommensurable biographies

If the epic body of the Apsara belies a simple answerability to gone-ness, the material bodies and biographies of the dancers themselves complicate this further. To be sure, not all Khmer Rouge survivor artists face these complications equally. In Los Angeles, to take only one example, Sophiline Cheam Shapiro has devoted herself tirelessly to the preservation and public performance of classical dance; her company Dance Celeste and her Arts of the Apsara foundation and school are a testament to her success.[6] In the case of the Sems, however, these complications were profound.

Even by the fraught standards of survival crises and refugee status, Ben Sem seemed to me to be a particularly damaged person. He drank heavily, though not with his contemporaries in Long Beach's Khmer neighborhoods but at a relatively out of the way bar frequented almost exclusively by Mexican and Mexican American laborers. He was easily angered and would fly into rages, shrieking and pounding the walls or the furniture with his fists. While I never saw or heard any evidence of him striking members of the family, it was clear that his rages were a kind of affective tyranny permeating the household. May intimated that they had lost friends due to these outbursts. Indeed, though I was told they rehearsed and performed for friends, I did not see a single adult visitor in the apartment in the years I knew the family which, based on what I knew about the affective and logistical interdependence in refugee communities, struck me as strange.

There were other Khmer dancers in the area, as well as active cultural organizations. When I asked Ben about them, he was frankly contemptuous: 'Easy say, over here, "I am dancer." In camp, some guys, all the time, "Oh I was dancer. I am actor." My kids say too, "Oh I learn the dances." Then "Oh, go with friends now." It is more, not just to think or say, "I am dancer." I say, "Okay. Dance."' The implication was clearly that these others were inferior, if not frauds. He, on the other hand, and in his son's words, danced 'real Cambodian dance.'

Certainly part of Ben's frustration was that the art which, in his mind, so clearly legitimated his survival would not, it appeared, insure continuity across generations. Rith, Sandy and Jennie simply were not as invested in classical dance as their parents. Sandy observed: 'I like to learn it [the dance] because it's important to know my culture. I know. But it's different for me than for them [her parents]. For them, it means more. I wasn't there. It's hard to relate.' For Jennie, the youngest, the ghosts that urged her mother to 'do the step better' seemed more like roommates: not especially compelling and sometimes a nuisance.

> *Jennie*: I know these monsters [ogres and trolls] and monkeys and
> stuff [all figures from the repertory] aren't real, but I feel like I
> know them because we see them all the time. I think my mom and
> especially my dad – I think they think they're real. I think they
> wish they were. Sometimes it's like they live here.
> *JH*: What do you think about that?
> *Jennie*: Um . . . It's okay, but sometimes they take up a lot of room.

May confided to me that Rith, who had been the most ardent dancer of the three children, doing cartwheels in the apartment to show off his fitness for monkey roles, was not coming home for multiple nights at a time, and that he had taken up with older 'dirty' boys, 'dirty' being her devil term for everything from the vaguely disreputable to the murderous. In what I can honestly say was the only time Ben ever clearly expressed vulnerability to me, it was to acknowledge this break in the generational chain of answerability he seemed to want so desperately: 'I am so tired some day. Why I fight with my children, "No TV, no out with friends, no this loud music." Why so hard to teach this dance? It live to come here, die to stay here? Sometime I think yes. Then I am so tired.' Certainly this break in the continuity of answerability across generations was made all the more painfully visible by my presence, an unmistakable indicator that the pure audience for the pure product of Khmer culture was polluted, corrupted by the reality of somewhere else and someone other. My interest in the dance, over and against his children's relative disinterest, could probably only be, for Ben, an index of his failure.

And yet it seemed to me there was something more than cross-generational frustration and corrupt witnesses at work in Ben Sem's performances of answerability. He made many pronouncements about dance; these comprised the bulk of our interactions. He did exercises repeatedly. Yet, in spite of his admonition 'okay, dance,' he actually danced very little, especially compared to May. This surprised me as May said she had studied only briefly at the Conservatory in Phnom Penh, and continued her training under the tutelage of fellow dancers in camp. Though I was never given dates, I understood that Ben was at the Conservatory much longer and that May thought he was 'good student.'

When I asked May why Ben didn't dance more, she paused and told me things which didn't seem, on the surface, to answer my question but which had clear implications for the precarious palimpsest of answerability to goneness that they had constructed together, and for the thresholds of silence which both maintained and undermined it. She was adamant that I not write down a single word. What follows is paraphrase.

May met Ben in a refugee camp English class, which she was taking to improve her chances for resettlement, to alleviate the boredom so pervasive in the camp, and to crowd out the memories of death, hunger and forced labor in the killing fields. Ben was older, much older I suspected. He impressed her because he was the best student in the class and, though he clearly separated himself from the other students, he was nice to her. He confided that he too was a dancer, but he didn't want to make a show of it. May was attracted to him because he was smart, because he needed her, and because, I suspect, her options were relatively few as a whole generation of men were casualties, not only of the Khmer Rouge, but also of the previous civil war. They began to spend time together, practicing English and dance exercises. He told her she was a beautiful Apsara.

One day, the distant relative with whom May was living in camp told her that some people stopped by, concerned about her association with Ben. The gist of their concern was that he associated with 'bad people' back in Cambodia and that he didn't belong in this camp but in another with his friends. May did not mention the name of this other camp, which would have been revealing as various Cambodian factions, including members of the Khmer Rouge, were separated, each to one of eight locations.

May told me that she thought these people were, in essence, jealous of Ben's intelligence, going so far as to describe them as thinking like the Khmer Rouge in their mistrust of a smart man. Yet, she admitted, she was concerned enough to consult a seer. The seer told May that Ben was not what he seemed – I do not have her exact words on this point – and that he had had another wife in Cambodia. May assumed this meant that Ben was not single, as he appeared, but why judge someone harshly for losing, or losing track of, a spouse when this happened to so many, indeed when she could not find one member of her own immediate family? May concluded by repeating a point she made often: Ben had seen many bad things. She added that, sometimes when he screamed in his sleep, she thought about what the seer said.

In her introduction to *Trauma: Explorations in Memory*, Cathy Caruth notes that 'what trauma has to tell us – the historical and personal truth it transmits – is intricately bound up with its refusal of historical boundaries; . . . its truth is bound up with its crises of truth' (1995: 8); this seems equally relevant to Apsara bodies, which transcend historical boundaries, and to material ones which do not. I am not able, and have no desire, to determine precisely how Ben Sem was not what he seemed, or which bad people he associated with, or if, indeed, there is validity to either of these assertions. It was clear from May's response to my question, though, that perhaps, in complex, painful and unknowable ways, the 'guilt and liability to blame' inherent in answerability might conjure ghosts independent of, and antithetical to, the monkeys, ogres and trolls of the *Lakhon Khol*, and to the transcendent, celestial Apsara, as Ben tried to inhabit and impart them. Perhaps, in the final turn, for Ben the palimpsest of answerability I have described, already precarious, became a chasm between two impossible bodies: one ideal and redemptive, the other material, historical, answerable to both utopian voices and darker sounds.

Like the epic bodies in Khmer dance, refugee exigencies are inflected differently by gender. It was not clear to me that May 'knew,' in any simple, denotative way, the 'truth' about Ben's past. I did not know if she explicitly reckoned with the possible incommensurability between that past, 'everywhere punched and torn by ellipses' (de Certeau, 1984: 107), and the smooth surfaces of Khmer repertory Ben deployed to fill in those gaps. It was apparent, though, that May had been disproportionately charged with the interpersonal and social responsibilities for managing both tactical expressions and strategic silences.

In 'Migrant Identities: Personal Memory and the Construction of Selfhood,' Keya Ganguly asserts that recollections of the past serve as ideological terrain on which people strive to represent themselves to themselves and, further, that this past 'acquires a more marked salience for subjects for whom categories of the present have been made unusually unstable or unpredictable' (1992: 29). Ganguly's formulation is useful for examining the multiple perils and labors facing May Sem – gendered perils and labors which cut across both past and present.

Material exigencies were acute and wearing. The Sems continually and unrelentingly negotiated the privations of refugee life. They had no car and, for some of the time I knew them, no phone. It was clear that much of Ben's tolerance for my early visits was attributable to the groceries I brought with me under the pretext of not leaving them in my car while I 'dropped in' and then 'forgetting' them on my way out. The need for the groceries became clear to me during my second visit, when I noticed the Sems's larder on that day consisted of a bag of rice, some dark leafy greens clearly past their prime, a battered box of sugary children's cereal, some loose tea, and two half-filled jugs of orange-flavored drink. The need for the pretext was established less directly, through both Ben's and May's prideful and plaintive characterizations of their 'work hard, not like dirty people in America lazy at home and drink the beer.'

Yet, just as with Ben's proclamation 'Okay, dance', these assertions of 'work hard' seemed poignantly problematic in deeply gendered ways. The Sems were 'partners' in a local donut shop though, as noted earlier, their use of the term was not, as far as I could determine, a formal, legal one. Donut shops were popular entry businesses for this first wave of Cambodian refugees arriving in Long Beach in the early to mid-1980s (see Pearlstone, 1990: 84). I understood that both Ben and May worked at the shop, and that the children sometimes worked there too. The arrangement, as I understood it, was that Ben worked during the day when the physical demands of the job were the greatest, and May took the equivalent of the night shift. In my experience, though, Ben was usually home during the day; occasionally, when he would leave in the middle of one of my visits, he and May would exchange words which, I learned later from the girls, centered on his not going to work. 'The big boss was mad' at him for absenteeism and 'yelling at people,' Sandy confided. May, on the other hand, went to the shop at her assigned time. I drove her when I was at the apartment; most days she took the bus. When the children were in school, this schedule was oppressive enough but, during the summers, burdened with concerns about her son and charged with managing the relative confinement of her daughters during the day, this late-night shift must have been unbearably exhausting. Certainly the physical fatigue was compounded by a maternal, relational one. I was acutely aware of May's anxiety and frustration, particularly when she had to leave for work and Ben was not yet home to watch the girls. Though very spirited, the girls' fears at being left alone at night in what could generously be described as a challenging neighborhood was palpable and, I saw, profoundly burdensome for May.

In addition to shouldering all of the domestic responsibilities and her shift at the donut shop, May sometimes worked Saturday afternoons cleaning at a local beauty salon. When she said she 'clean the hair,' I first thought she meant she shampooed customers – a path, I hoped, that might lead to an entry level cosmetology job or, eventually, even a license. I learned later that she meant 'clean up the hair.' Unfortunately, May eventually had to give up this job after she came into contact with some chemicals that hideously blistered her hands. Ben was furious; the 'big boss' would not let May work with or around food as long as she had open sores, so two sources of income were, at least temporarily, lost. Still, Ben would take money – sometimes a few dollars, sometimes change, to go the bar to 'drink the beer,' or to wander. Rarely, new Khmer music tapes would appear. Rith came home one day with new, expensive gym shoes that led to a fight with his parents and ended with May in tears. They did not buy the shoes, she explained later, and he wouldn't say where or how he got them.

In Khmer culture, gender roles are both fixed and fluid. Theoretically, men have primary responsibilities outside the home and women inside it; in practical terms, these divisions meant little in agricultural families and, in urban settings, women often set up and ran side-businesses (see Welaratna, 1993: 30–2, 233–4; and also Becker and Bevene). Yet the massive, and successive, psychic and physical dislocations suffered by the Khmer refugees in the United States, coupled with widespread unemployment in Southern California during the early to mid-1990s, left many men unable to find jobs paying a living wage, families even more fractured, and women like May profoundly and multiply burdened. Shoring up the family's material survival was only the beginning.

In *Beyond the Killing Fields: Voices of Nine Cambodian Survivors in America*, Usha Welaratna interviews 'Apsara,' a pseudonym used for a first wave refugee woman living in northern California. Like May, 'Apsara' had training as a dancer, but here the similarities end. Welaratna describes 'the laughter and chatter' in the household, the relatives and friends visiting, the 'strikingly Western' look of her self-directed interlocutor at her workplace (Welaratna, 1993: 219, 220). This was a very different world from the Sem apartment, which was generally quiet except for Khmer music on a small boom box, noise from the television, or both. No friends or relatives appeared or were mentioned; if a call had to be made during the periods in which the family lacked phone service, Sandy was given the information and sent to a neighbor to borrow the phone. May was clearly quite isolated; once, when she seemed on the verge of tears, I asked her if she had friends to talk to. She replied: 'Husband say not [gestures out the window, then inside] and not in to out.' I learned later that this was, perhaps, a version of a Khmer proverb that admonishes husbands and wives: 'Don't bring the fire from the outside inside the home' (Sovann Tith quoted in J. Stewart, 1997). Indeed it was clear from Ben's thorough reluctance to share even the most denotative biographical details, and May's adherence to this, that 'not in to out' was a discursive category both broad and deep.

As noted above, my visits and questions about dance exposed deep fissures in generational continuity for Ben Sem. As such, I suspected that he found them at least somewhat taxing. May, though, seemed to welcome them. She said as much. Above and beyond the materiality of rides and groceries, I suspected she enjoyed the company. Though the image of a cultural outsider taking up what should have been her children's interest may have crossed her mind, my outsider status was also, I think, a relief. I didn't speak Khmer and didn't live in the community, so I couldn't pass on gossip to the neighbors. She could 'learn more English' from someone other than her daughters, who were quite fluent and corrected her, to her evident embarrassment.[7] Most of all, I was willing, even eager, to collude in May's desire to use dance as an alibi – as a code for talking about and around things that could not otherwise be said:

> *May*: I like when we talk about the dance but I don't like other talk.
> *JH*: The other talk?
> *May*: When you want to talk about here and about camp. I don't like. We don't talk that time. I think we talk about the dance so you know everything that way. Understand? Okay? We talk about the dance so then you know.

While Khmer classical dance was a code that organized my relationship to May, her relationships with her children seemed always fixed at the precarious intersection of dance and the 'other talk.' About a

year and a half into my visits, it became clear that Rith's primary interpersonal investments lay outside the family. In an irony not lost on
his sisters, the gentle, outgoing boy who was acutely affected by his
father's rages became increasingly tight-lipped and short-tempered; he
was away from home frequently, particularly on weekends, and, when
pressed, he was given to slamming doors or pounding on walls. Though,
as I have noted above, I did not speak or understand Khmer, it was
clear to me that both Rith's tone and his nonverbal demeanor toward
his mother was becoming increasingly dismissive and harsh. May was
keenly aware of her son slipping away from her. Her assertions that
Rith 'is so good boy, so good dancer' became less and less frequent.
Toward the end of our relationship, she did not speak of him much
at all.

Perhaps this sense of her son slipping away contributed to her tight
hold on her girls. Khmer norms of propriety for young women were
certainly a factor too. They were almost completely confined to the
apartment after school, on weekends, and during the summer. 'In
America so many dirty boys trick the girls go with them,' May admonished her daughters when they pleaded to go out with friends from
school (see also Welaratna, 1993: 276). 'You stay home, learn the dance.'
Needless to say, the girls' execution of dance exercises ranged from the
half-hearted to the sullen after these exchanges.

If May negotiated the distances between herself and her children with
rigor or resignation, depending on gender, Ben's responses were predict-
ably volatile. This, in turn, generated even more emotional, relational
labor for May as she attempted to mitigate the affective fallout of his
rages. As I mentioned earlier, I never saw Ben strike a family member;
indeed I became more worried about Rith in this regard. I explicitly asked
Sandy about physical discipline in the home and she replied emphatically that, unlike her friends' parents, May and Ben never hit their
children.[8] While this was a relief to me, it was clear that Ben's anger and
his own interpersonal damage was corrosive and needed to be managed.
It fell to May to take on a relational, as well as a material, second shift.

During one of my visits, Ben was particularly agitated by the children's bickering. They were fighting over the television; its volume,
and theirs, fluctuated wildly, drowning out the Khmer music Ben was
listening to, his head rocking slowly from side to side as he stared out
the kitchen window. He began to scream angrily and seemingly inco-
herently – shrieking, really – and then stormed out of the apartment.
After tending to the children, May took me aside. This was one of only

a handful of times she explicitly acknowledged her resentment of the relational labor she was forced, by extremity, to perform:

> *May*: Sorry, you know. [I am] sorry for him because I think noisy [talk, voices] like camp and hard for him. Hard too for me. Hard for me too, but I can't [mimics Ben storming out]. I have [points to her daughters]. *I* have to tell them why. *I* have to say, 'Don't talk. Don't talk about that. Don't be so noisy.' *I* have to say, 'When you are bigger, you will know he see very bad things.' I see bad things too, over there. It is bad for me too, over there. Worse. Worse, I think, maybe, but I can't say. I see bad things too but I [shakes head 'no'] say and I [shakes head 'no' again and mimes storming out] because of them [her children].

May saw quite clearly that her present was precarious because her and her husband's pasts were even more so. Leaders of the Khmer community in Long Beach knew this clearly too. As Chetra Keo of the United Cambodian Community, a grass-roots services organization in Long Beach, observed: 'The whole population has been abused by the Khmer Rouge regime... Culturally speaking, we come from a Buddhist practice. We are not taught to be aggressive. But because we went through war, through the Khmer Rouge, and the refugee camp in Thailand, there were so many years of violence that we saw before we came to this country' (in J. Stewart, 1997: B3). For many in the community, the personal past, told as personal narratives, served as ballast to support the construction of post-Khmer Rouge selves, and as antidotes to the Khmer Rouge's enforced cultural amnesia. These are stories of survival (see LaFreniere, 2000; Pran, 1997; Ung, 2000; Welaratna, 1993; among others), even when permeated by grief (as in Ngor, 1987). Yet for Ben and May, the personal past was as fraught with instabilities as the present. No coherent story would, perhaps could, be told.

As a wife and mother, May was charged with keeping material, affective and discursive order in the household. Yet at the same time, she lacked the power to establish the order she was impelled to maintain. Ben's injunctions to remain silent were central to that order, at least as far as I could determine. May had to generate familial continuity despite his insistences on crucial omissions. With the representational terrain of both past and present – 'the other talk' – too perilous to deploy, May, like her husband, turned away from both in favor of an option more utopian, more remote. To return to Ganguly's formulation, the transcendent past of the Apsara, *not* recollections of the personal past, became the ideological, representational terrain for May's self-fashioning. In so doing, May, like her husband, faced unique, and uniquely gendered, incommensurabilities between her life, her history and her art.

At the most denotative level, May's attempts to deploy the Apsara to answer back to goneness, displacement and privation ran up against the limits of her training. She had never been a performer with the Court Ballet, only a student at the Conservatory and, in her words, 'not good' student. Though May was deeply guarded about details like names and dates, she did reveal the general outline of her dance education in a fairly detailed account, which I quote at length:

May: [holding up one finger] One sister [first sister/eldest sister] go with me – see old teacher [*an* old teacher, not her sister's old teacher]. Old teacher know my mother long time. Mother say, 'This one always dancing. You go see her.' So One sister take me. Teacher ask me the foot, the hand, walk, like that. Say maybe I go to [dance] school, but maybe too big girl. So I want to go! I want to go! 'Mother, mother, tell old teacher!' On and on, everyday, like that. My father say, 'Go tell teacher or we go crazy!' But mother say I am not so good girl. I am noisy, talk too much. Not so good. But father say to One sister, 'You go with her. Take her to the school.' So I go!

But school is so hard place. The hands hurt, the [thighs] hurt. Go like this [mimes manipulating the shoulders]. Boo-boo-boo [mimes throbbing], like that, all night so no sleep. After the dance, then to the other school. Hit with stick for noisy.

I go. Love the dance but I am not so good student like some so pretty girl. But I am (finger on the table, drawing a line, presumably 'straight') and so small [thin, not short; May was considered tall as a girl] and I do with music like that, so good, so I stay. Girl go, girl go, girl go. I stay. Teacher say, 'May, you good student you get the money for your family so be good girl. Not so [mimes bouncing around, the equivalent of 'all over the place'].' So come big dance for some big bosses – not the King but big people, like that. And teacher say only so good girls, so pretty girls dance. I am too big girl. Girls [mime pointing] 'Maybe May do the monkey. Maybe May do the monster.' Like that. I say, 'You fat girl, you fat foot.'

My father say, 'May, you close the mouth.' He so good man, he say, 'Dance need the big girl too, not only the so small girl. Another dance for you coming.' In camp, I think about this.

Her father's sense of 'another dance' coming for May might have been meant literally. As Toni Samantha Phim and Ashley Thompson note in *Cambodian Dance*, 'girls with a smaller build and rounder faces will study the female roles; those with larger frames [like May, who was considered tall] and longer faces will play the male and ogre [monster] roles' (1999: 46). Yet it was hard to avoid thinking of this observation as both metaphorical and prophetic. Surely another dance, far removed from the serenity of the Apsara even if it shared the same corporeal infrastructure, did come for May Sem.

As May herself observed, even early on the Apsara ideal was an uneasy fit. This was not only a matter of her physicality. Though I don't have her exact words on this point, May intimated that she also felt isolated from the 'so pretty' girls in the Conservatory, some of whom came from 'big' (important, not necessarily large) families, and who made fun of her when she was passed over for the important performance she described. Though she was quite circumspect about her own family, I gathered from what I did learn that her father was a lower-level civil servant or clerk. While the family was not materially deprived, they were not 'big people' either. The fact that May might have been eligible for a stipend mattered, both to her and to her family. To call May's unease with these social aspects of the Apsara ideal 'class shame' might be overreaching. That said, this unease can also be viewed as a reflection of a broader ambivalence undergirding the sociality of Khmer classical dance.

The class politics of the Apsara ideal are more complex than they might at first appear. Historically, Cambodian classical dance, and the dancers themselves, 'were core elements in the [Khmer] ruler's symbolic display of prowess' (Phim and Thompson, 1999: 39). Performances and performers did not exist apart from royal patronage. Unlike theatrical and folk dances, or the all-male *Lakhon Khol*, which were grounded to varying degrees in village life, the Apsara ideal in performance was never a vernacular form.[9] Classical dance is court dance. Cambodian royalty has always been intimately involved, not only as primary supporters and producers of classical performances but in choreography as well. Consider two especially important examples, both involving the family of the twentieth-century ruler Norodom Sihanouk.[10] His mother, Queen Kossamak, is credited with shaping the dances to fit modern performance conventions, as well as adapting standard movements into 'new' dances. Sihanouk's daughter, Bopha Devi, herself an Apsara dancer, was central to the reinvigoration of the classical tradition in the post-Khmer Rouge Cambodia. Phim and Thompson note that she was named Minister of Culture and Fine Arts in 1999 (1999: 44).

Yet the material lot of the dancers themselves was of a much different order. Judith Coburn observes that, in addition to serving as spiritual intermediaries and cultural exemplars: '[t]raditionally [female] Cambodian dancers also had more earthly duties: to mate with the king.... Originally, dancers were recruited from villages all over the kingdom when they were about 6 years old and lived in seclusion in the royal palace for the rest of their lives' (1993: 16). These were not lives of quiet contemplation but of intense physical labor: 12-hour days of training, enforced by corporal discipline, were the rule (Coburn, 1993: 16; see also Mydans, 2001). Phim and Thompson observe that only 'favorites' accumulated any personal wealth (1999: 40). Moreover, '[m]ost of the women lived and ate communally. Meager rations had them anticipating visits from family who would bring extra provisions. Though admired, even venerated during performance, with the exception of those of royal lineage or those selected as a principal concubine, the dancers were not privileged members of society' (Phim and Thompson, 1999: 40). May, it seemed, negotiated the labors and privations of the Apsara without even the veneer of luxury that performance might have provided. The gendered social contradictions always already embedded in the Apsara ideal did not disappear when May deployed dance to answer back to goneness. On the contrary, the distance between the transcendent serenity of the form and the material circumstances of its (re-)production in a poor and traumatized refugee family widened and deepened.

In her sparsely furnished, dark living room, exhausted from manual and relational labors very different from those instantiating the Apsara, May moved from one simple pose to another. The multiple past and present displacements separating her from the dance she loved must have been wearing. I vividly remember her practicing hand gestures, flowing from one curved finger pointing upward, into four fingers doing the same, into a turn of the wrist with thumb touching the index finger and the other three fingers curving outward like the petals of a flower. This sequence more than any other seemed to be May's ritual of reassurance, a deployment of fragments of her training, however limited, and transcendence, however qualified by material exigency. She would turn to this sequence after moments of tension in the household, or when the girls rebelled against her injunctions to stretch or practice poses in favor of the TV. There was certainly a profoundly lyrical beauty in these spare and elegant gestures, but what struck me most forcefully was how May regarded her hands as she executed each movement. Her face was blank. This was not the sweet, otherworldly gaze of the Apsara, with its gentle half-smile. It was not an expression of concentration, but rather of a sad and detached weariness.

Years later, I saw what I believed was this same sequence of hand gestures again. On stage at the Japan America Theatre in Los Angeles ('The Horse's Mouth Greets the New Millennium,' July 22, 2000), Sophiline Shapiro narrated each movement, dissolving into the next, as an allegory of the re-emergence of Khmer culture. The gestures themselves represented a bud bursting into blossom, then into a fruit, eventually to reseed the arts in a generative cycle. The difference in tone between this performance and May's 'ritual of reassurance' was certainly stark. Yet another distinction struck me, one more denotative and technical. Sophiline's long, slender, elegant fingers curved backward from her lower knuckles, appearing at an almost 40° angle from her palms. This extent of flexed curvature was extraordinary to see; it could only be achieved through years and years of forcing the fingers back from the palms, beginning at a very young age. I knew the physical pedagogy; May continually used one hand to push back the fingers of the other and urged her girls to do the same. 'Beautiful hand not easy,' she would say. Yet I realized, seeing the performance of Sophiline, that the limits of May's training, her age, and the residues of trauma and hard work conspired to keep the classic 'beautiful hand' out of reach. Certainly May herself must have been aware of this. Further, I believe, in retrospect, that her persistent urging of these stretching exercises on her daughters must have been founded, at least in part, by the awareness that their time for 'beautiful hand' was running out.

I wondered, too, if this sense of technical and material limits, exacerbated by the passages of time and hardship, contributed to May's abiding by, one could even say colluding with, Ben's isolationist demands. I attributed Ben's refusal to perform in public to the unknowable particulars of his damage, and had assumed that May went along to placate him. The dynamics, I now suspect, were more complex. Perhaps May could not risk exposing the technical limits of her answerability to goneness, the cracks in the enabling, smooth surface of the dance that, in her mind, legitimated her survival. The risk, I suspect, was less about public failure than about deeply inter- and intra-personal issues. If dance was the technology Ben and May deployed to survive and answer for it, what were the consequences for their marriage, for the rituals of reassurance which at least nominally organized their family, and for their psyches if that answer was meager or faulty?

I wondered too if this explained Ben's and May's reluctance to attend performances. The Cambodian community in Long Beach supports a rich cultural life; though there were fewer opportunities in the early 1990s than there are today, the Sems certainly had performance options available. Yet Ben and May did not attend these events, not even the 1990 performance of the Classical Dance Company of Cambodia that was part of the widely publicized Los Angeles Festival. Rith attended in the company of those I assumed were relatives (I later had reason to believe I was wrong about this), as did a sizeable number of the local Cambodian community. Perhaps Ben and May were working. Yet May's evasive responses to my questions about attending performances with me, or my taking the children to attend (I knew better than to ask Ben), suggested an additional possibility.

Answerability to goneness through performance cuts across many accounts of Khmer survivors. Daran Kravanh's story is illustrative: 'I cannot tell you how or why I survived; I do not know myself. It is like this: love and music and memory and invisible hands, and something that comes out of the society of the living and the dead, for which there are no words' (LaFreniere, 2000: 3). Kravanh describes a remarkable moment when, exposed by his compulsion to play his accordion, despite warnings that this marked him for death, he used music to bind him in solidarity to his Khmer Rouge would-be executioner:

If I was going to die, I wanted to die playing my accordion. . . . I played that song with all the joy I had ever felt. The soldier did not shoot me. He listened until the end of the song. Then he said in a small, quiet voice, 'Will you teach me to play?'
'Yes,' I said. Yes was all I had left to say.

(LaFreniere, 2000: 152)

Likewise, after the collective trauma of the Khmer Rouge, coalitions of performers mobilized to use performance technique as an affective and social infrastructure for healing and renewal (see, for example, Charlé, 2001; Coburn, 1993; Mydans, 1993; Turnbull, 2001). This is answerability in and as community, one of the 'somethings' to come out of the 'society of the living and the dead' that Daran Kravanh described. Yet more than one thing may emerge from such a society. As Kai Erickson observes in 'Notes on Trauma and Community,' 'trauma has both centripetal and centrifugal tendencies' (1995: 186). The Sems clearly occupied the centrifugal position on this continuum. The answerability they constructed through performance explicitly excluded memory ('We don't talk that time.') as well as any probing of technical limits that participation in community might expose, encourage or, indeed, require of both performers and audiences. Perhaps the complex sociality of spectatorship, with its multiple invitations to vicariously inhabit other bodies, other histories and other places, was simply too painful for Ben and May to endure.

Françoise Lionnet observes that the postcolonial subject 'becomes quite adept at braiding all the traditions at its disposal, using the fragments that constitute it in order to participate fully in the dynamic process of transformation' (1995: 5).[11] Lionnet's argument is specifically applied to postcolonial women writers, but what of subjects like May Sem, braiding together trauma, relational obligation, art and hardship – fragments which are themselves fragmented, and which are so shot through with silence and damage that they do not or cannot cohere? Much has been made of border zones where categories overlap and blur in the process of postcolonial and refugee self-fashioning but, in the case of this family, borders marked chasms, not confluences. If the multiple border zones of gendered dislocation offer opportunities for creative braiding, they just as surely expose the incommensurability of fragments of identities. These zones of dislocation offer possibilities of dismemberment, as well as of *metissage*.

In her presentation, 'Finding Pleasure in Margaret,' Della Pollock explores the 'monstrous maternity' she sees circulating in the film *Margaret's Museum*. In the film, the protagonist, Margaret, dismembers the bodies of her father, brother and husband, victims of mining disasters, and displays 'the carefully cut body parts... bottled, labeled, and preserved for show' (2000: 4). Pollock's analysis of monstrous maternity as a trope in this film resonates with May's deployment of dance to answer for her survival. Like Margaret, May is 'the curator of her own home,' (Pollock, 2000: 4) not literally, but certainly both metaphorically and explicitly ('So Khmer dance don't die too.'). Like Margaret's museum, May's construction of answerability 'showcases loss. Past and future losses that carry no trace value,' that, indeed, cannot support the sociality of evaluation (Pollock, 2000: 5). Pollock sees Margaret's 'desire and devastation magnify[-ing] each other to gigantic proportions, threatening to burst their containers' (2000: 5). May's trajectory seemed to me to be the reverse; desire and devastation, answerability and incommensurability, seemed to pile up on one another like stones, threatening, it seemed, to implode into some sort of black hole of exhaustion and silence. In *Margaret's Museum* and in May Sem's deployment of classical dance as answerability to goneness, Pollock reminds us, 'the possibility that remembering truly gigantic loss [and answering for it] may be less liable to our fantasy of incorporation than it is kin to dismembering or, in this case, to the cut up incoherence of' and in performance (2000: 6).

Ben and May were fragile people. In May or June, 1992, exactly when I don't know, they abruptly left the city for northern California. I learned this from neighbors, who knew no more. The Sems themselves never contacted me and my attempts to contact them met with no success.

Bruno Bettleheim once observed that psychoanalysis explains why we go crazy, not why we survive (*How Nice*). Despite Ben's and May's insistence to the contrary, the daily rituals and invocations of performance didn't appear to explain it either. Perhaps this is because this 'why' is often permeated by exhaustion, by crises of truth and thresholds of secrecy. For Ben and May Sem, saving Khmer classical dance as a daily legitimation of their survival offered both an explanation and solace and, at least in the time of our acquaintance, delivered neither.

4
Dancing Other-Wise: Ethics, Difference and Transcendence in Hae Kyung Lee and Dancers[1]

Transcendence is born of the intersubjective relation.
(Pierre Hayat, Preface, *Alterity and Transcendence*, 1999)

Codes and criteria are indispensable parts of ethics, and surely they will not work without a sense of obligation or subscription. But these last things are still not sufficient to the enactment of ethical aspirations, which requires bodily movements in space, mobilizations of heat and energy, a series of choreographed gestures, a distinctive assemblage of affective propulsions.
(Jane Bennett, *The Enchantment of Modern Life*, 2001)

Because we can share our mortal time and touch only with some and not all, presence becomes the closest thing there is to a guarantee across the chasm. In this we directly face the holiness and wretchedness of our finitude.
(John Durham Peters, *Speaking Into the Air*, 1999)

Dancing communities use technique to solve material problems in the here and now. Technique's protocols make the body communally readable and available. They offer interpersonal, social, and archival infrastructure. And they generate primary and secondary rhetorics of affiliation that bridge multiple dimensions of difference within the global city. Yet dancing communities, and technique itself, also organize and, in turn, are organized by the ineffable. For the Sem family in Chapter 3, the celestial Apsara proved too fragile to withstand the withering effects of horror and privation. In other dancing communities, though, technique anchors a collective metaphysics in the material. Here, dancers use rhetoric and images of transcendence to build solidarity and incarnate ethical opportunities and obligations to their audiences and to each other. Technique uses metaphysics and is, in turn, enchanted by it. If the former is, at the simplest level, denotative, making the body legible and communicable, then metaphysics, the latter, is another voice that offers a way of constituting a company community, of characterizing the transporting effects of legible bodies' beauty, and of grappling with the capriciousness of those bodies' finitude.

This chapter examines the productive intersections of transcendence, ethics and multiple dimensions of difference in an internationally renowned dancing community: Los Angeles-based Hae Kyung Lee and Dancers. Hae Kyung Lee was born in South Korea and began her dance career there. She has performed as a solo artist for over 20 years and has received wide international, national and local acclaim, including an NEA Choreography Fellowship and residencies at the Palucca Schule (Dresden, Germany) and the London Contemporary Dance School (United Kingdom), among many others. Her company, Hae Kyung Lee and Dancers, is now in its fourteenth season. Though company membership fluctuates, four dancers have formed a consistent corps/core over the last ten years: Claudia Lopez, her husband Miguel Olvera, Kishisa Ross, and Jill Yip were all former students of Hae's at California State University, Los Angeles (CSULA), where Hae is currently a Professor of Dance.

I met Hae when she joined the CSULA faculty in 1993. She seemed to personify the possibilities of dance in the global city: an artist with an international movement vocabulary and training, committed to building a multi-ethnic community of artists, and to exploring new creative possibilities. Her effect on the dance program was catalytic. I followed her and her company for ten years, attending rehearsals and performances, conducting interviews, and sharing key events in their lives and work. Three members of the company, Claudia, Miguel and Jill, were my students in undergraduate and graduate courses.

Hae's aesthetic draws from her Korean Buddhist heritage, her Western training in modern and postmodern dance vocabularies, and the exigencies in the lives and bodies of her dancers. Viewing the Company's dances is slightly disorienting in ways that parallel listening to members' conversations. In the latter, metaphysics breaks into everyday talk, recasting what seems like an ordinary conversation in ethical terms while leaving the mechanics of this recasting unspoken. In Hae's choreography, lyrical beauty seems to erupt within the most focused, deliberate and challenging demonstrations of precision and strength. The reverse is also true; moments of effortlessly elegant movement suddenly inspire a purely technical response: 'How do they **do** that?' Much of Hae's choreography involves holding such oppositional elements in tension without easy resolution.

This chapter sometimes speaks in two registers. It alternates between italicized passages presenting company life and discourse, and analysis of that discourse. This is not to suggest that conventions of academic analysis and daily talk within the company are opposed; rather, it is to give the reader a sense of how an observer would encounter the metaphysical formulations company members use as a matter of routine. These formulations have an opacity that is difficult to describe. They erupt within, and sometimes seem at odds with, the critical precision the dancers use in casual conversation.

'The universe's piece'

I am overlooking the Watercourt of the California Plaza on Grand Avenue in downtown Los Angeles. Dancers, wearing what appear to be short trashbag jumpsuits, are moving about in the rushing water on the highest level of a large, multi-tiered fountain/environment. This is a rehearsal for Ancient Mariners, choreographed by Hae Kyung Lee, performed by her company Hae Kyung Lee and Dancers, and commissioned by Grand Performances. The company began its eleventh season (2000) with this performance. In addition to the four core members of the company, Ancient Mariners featured Hae's original dance partner, James Kelly, CSULA dance students Henry Mendez, Paula Miyashiro and Jose Ramos, and LA dancer Derrick Jones.

Jill, a 'first generation company member,' is coaching Paula, a company newcomer. As the two of them work, I am free to talk with Hae. She tells me incredible things. Composer Steve Moshier completed the multiple-movement score, but she had choreographed the piece independent of the music, never hearing it in its entirety until she was finished. Her work and his are perfectly matched in tone and momentum in each and every movement. Further, she choreographed the piece, nearly an hour long, with nine dancers, multiple levels to the Watercourt, and multiple, changeable configurations of water flow, in only three days. I am incredulous. She is matter-of-fact. It is the universe's piece, she says.

Referent trouble and the metaphysical performative

To spend time with dancers is to enter conversations replete with the ineffable, with references to the transcendent, the metaphysical. Dancers speak of movement and performance as spiritual necessities, as healing or as solace. Choreographers describe states of possession or aesthetic arrest. They defer their own agency to that of the divine.

Metaphysical invocations are not limited to ballet, where the other-worldly are literally staged. Whatever justifiable claims can be made for modern dance as political, as constitutive of national identity, as feminist in its semiotics and contexts of production and reception (see Banes, 1998; Daly, 1995; Franko, 1995; Martin, 1998), references to the ineffable were central to the technique from its inception. Metaphysical pronouncements circulate through modern dance, as in Martha Graham's assertion that 'Movement never lies' (1998: 66). In her essay, 'I Am A Dancer,' she writes:

> Ambition is not enough; necessity is everything. It is through this that the legends of the soul's journey are retold with all their tragedy and their bitterness and sweetness of living. It is at this point that the sweep of life catches up with the mere personality of the performer, and, while the individual becomes greater, the personal becomes less personal. And there is grace. I mean grace resulting from faith... faith in life, in love, in people, in the act of dancing.   (1998: 67)

Ruth St Denis embodied Asian deities. Sally Banes describes Mary Wigman's accounts of 'possession' in the creation of her 'Witch Dance' (1998: 127). Isadora Duncan invoked Dionysian rapture. Choreographer Yvonne Ranier marked her departure from the expressionistic excesses of modern dance in the early 1960s, characterizing the break as a search for her 'own holy mission' (Banes, 1998: 219).

The issue here is not whether artists believe in the metaphysical discourse they invoke; they may or may not. Nor is this a matter of explicitly deploying metaphysics as a rehearsal or performance tool in the studio. Instead, the issue is how such invocations, such rhetorics of the ineffable, function, particularly when they are imbricated in the daily practices of dancing communities.[2] In the multi-ethnic, urban, contemporary dance company I describe here, invocations of the ineffable, the transcendent and the mystical performatively manage both alterity in community and in the creative process by providing a non-ontological ethics. These ethics are rooted in a generative notion of physical and social presence.

Both dance and metaphysics challenge referentiality. Perhaps this is why they are so often partners. Transcendence has always, ironically, required embodiment and generated its own technical protocols for engaging the body. This relation is especially acute in mystical discourse. As Michel de Certeau observes: 'because religious terminology could no longer be trusted... the mystics were drawn away, by the life they lived and by the situation that was given to them, toward a language of the body. In a new interplay between what they recognized internally and the part of their experience that was externally (socially) recognizable, mystics were led to create from this corporeal vocabulary the initial markers indicating the place in which they found themselves and the illumination they received' (1992: 15). But these markers were not simply indices to private, intrapersonal experience. Instead, they rendered the ineffable readable, communicable and social: 'The thread of psychosomatic signs was... the borderline that made it possible for mystical experience to be articulated in socially recognizable terms, to be made legible to the eyes of unbelievers. From this viewpoint, mysticism found its modern social language in the body...' (de Certeau, 1992: 15). Mysticism is a special limited case of the sociality and communicability of transcendence, yet its emphasis on corporeality is not unique. Emmanuel Levinas locates transcendence in the face of the other, in the other's irreducible corporeality. Here, transcendence turns on responsibility for the other in ways that are especially relevant to the ethical community I discuss below.[3]

Yet even as the body speaks in the special case of mystical discourse, or when characterizing the intersubjectivity of transcendence more generally, it also 'unsays'; it exposes the ineffable's resistance to language. Indeed a corporeal grammar of the spiritual is predicated on such resistance ('Because religious terminology could no longer be trusted'). This resistant relation also surfaces in dancers' accounts of movement. Sometimes this appears as a kind of sacred inarticulateness, in which language is banished entirely, or dismissed as the corrupt partner of a purer, more genuine mode ('Movement never lies'). At other times, it emerges in the process of rendering dance discursive, fixing the moving body in language and, particularly, in writing, as Susan Leigh Foster observes: *'Am I pinning the movement down, trapping it, through this search for words to attach to it?... We thought any attempt to specify more than dates, places and names would result in mutilation or even desecration of the body's movement. We gave ourselves over to romantic eulogies of the body's evanescence, the ephemerality of its existence, and we reveled in the fantasy of its absolute untranslatability'* (1995: 9; emphasis in original). The dancing body, and the mystic one, explicitly identify the limits of representation in/as translation. Both resist conventions of referentiality even as they reveal that these conventions can never be simple. Michael Sells characterizes this 'dilemma of transcendence' denotatively: 'The transcendent must be beyond names, ineffable. In order to claim that the transcendent is beyond names, however, I must give it a name, "the transcendent"' (1994: 2). De Certeau presents this resistant referentiality of the ineffable more lyrically, in terms that couple discursive paradox with longing: 'What should be there is missing. Quietly, almost painlessly, this discovery takes effect. It afflicts us in a region we cannot identify, as if we had been stricken by the separation long before realizing it' (1992: 1).

De Certeau writes of the discourse exiled from the reassuring conventions of referentiality,[4] but corporeality may also share this banishment. Elaine Scarry argues, for example, that pain undoes language: 'Hearing and touch are of objects outside the boundaries of the body as desire is desire of x, fear is fear of y, hunger is hunger for z; but pain is not "of" or "for" anything – it is itself alone. This objectlessness, the complete absence of referential content, almost prevents it from being rendered in language' (1985: 162). Pain's dance with referentiality is complex. Like mysticism, it is a limit case, marking the always irresolvable dilemma at the center of relationships between language, the profoundly intrapersonal, and the intersubjective. The urgency and necessity of assertion is always coupled with the impossibility of speaking 'truly' or 'accurately.'

Dilemmas of transcendence and of corporeality arise as dancers try to characterize both; interestingly, the formulations used to express the spiritual in performance, perhaps the ecstasy of success, may be exactly the same as those used to express the pain of injury: 'I can't explain. Something took me over. It's beyond words.' While there is no way to resolve these dilemmas, dancers do manage them using a variety of tactics.[5] Technique offers one denotative possibility: viewed as protocols of reading and writing the body, it allows the social experience of that body to be named with relative precision by artists and spectators, even if the interior, phenomenological experience of the dancer herself is 'beyond words.'

Another option is silence, the 'sacred inarticulateness' noted above. One common variant is to opt out of language in favor of an 'articulate body,' a move that risks reinscribing a language–movement binarism which, in the day-to-day sociality of dance, is unworkable. Barbara Browning observes: 'Our frequent admonition – stop thinking and dance – isn't to say that motion is unthinkable. It's to say that the body is capable of understanding more things at once than can be articulated in language. One has no choice but to *think with the body*' (1995: 13; emphasis in original). But thinking with the body, like pain or spiritual experience, isn't pre- or a-discursive in any simple way. As Della Pollock observes, pain – and, I would add, both corporeality and ineffability – circulate through stories which expose it as 'interwoven with meanings and meaning systems' (1999: 121), systems which, like dance technique, insert both the body and metaphysics into the lexical and the social, however uneasily.

Stories and ritual performances manage dilemmas of referentiality involving both the body and the ineffable. Here, as Michael Sells observes, the dilemma 'is accepted as a genuine *aporia*, that is, as unresolvable; but this acceptance, instead of leading to silence, leads to a new mode of discourse' (1994: 2). In dance communities where rhetoric of the ineffable meets the labor of the corporeal, this new mode generates, not a 'thinking body,' but 'mobilizing fictions' of the transcendent that defer denotation and resist referentiality even more explicitly. Metaphysical vocabularies both point to linguistic, intrapersonal and social lacunae in dance and solve them. These vocabularies, these mobilizing fictions, like technique itself, are tactical tools for bridging referential gaps, and affective and social ones, that constitute the daily communicative life of a dance company. These gaps include those separating individual agency and its effacement in group solidarity; they include the divide between the alterity of creation and alterity in community, between inserting the body into technique and exceeding it, between, as Peters says, 'the holiness and wretchedness of our finitude' (1999: 271).

While technique offers a solution to the referent troubles of describing the body in motion, one difficulty with weaving metaphysical vocabularies into a community's mobilizing fictions remains: resistance to denotative fixity. A symptom can be found in my own prose here: the proliferation of roughly equivalent, though not completely synonymous formulations like 'the ineffable,' 'the spiritual,' 'the transcendent,' 'metaphysical.' Yet, in another way, this proliferation is highly productive; it evades not only strict referentiality, but also theological fixity. Slippage among these formulations is productive and useful because it so readily accommodates a variety of readings, a variety that is inevitable in a highly diverse company like Hae Kyung Lee and Dancers. Indeed the very utility of metaphysics as mobilizing fiction depends on this referential instability. It can insert the unruliness of creativity in dance into the unruliness of representation without defining, fixing or regulating either one.

In Hae Kyung Lee and Dancers, metaphysical vocabularies are not offered as explicit grounds for reflection, as Zarrilli suggests in his 'metaphysical studio' (2002: 157); there is no pedagogy of paradoxical questions to be explored, no emphasis on any formal ritual exercises like breathing. Instead, the ineffable is routinely invoked in everyday talk. It is the background to the company's training and sociality, not the foreground. These formulations do not arise from communally sanctioned formal or informal sets of rules, organizing principles, or definitions for terms like 'the universe,' or 'spirit,' or, as will be discussed below, 'energy.' There is no uniform regime of belief that anchors references to the transcendent for everyone in the community.

Such references are, in effect, metaphysical performatives that create and stabilize communal investment, not necessarily belief, with their utterances and repetitions. The rhetorical utility of such formulations coincides with their use. Metaphysical performatives in this company are tactical. Because every member of the group understands them differently, and because no essential definitions are ever offered, individuals enter, witness and use invocations of the transcendent from different cognitive and affective places. Some see them as an 'in-group' discourse, a kind of jargon; some as Hae's idiosyncratic pedagogy; others enfold them actively into their spiritual lives. But there is no prescription for doing so and no proscription against disregarding them entirely. To put this in religious language, it is impossible to 'blaspheme' a metaphysical performative in this community. It has no pure, pristine form to which all ascribe, no stable referentiality that everyone assumes or shares. This heterogeneity, this simultaneity of contrary possibilities for entering in and understanding metaphysical performatives, is, however, harmonizing in its effect. The productive vagueness of these formulations makes them a generative rubric for uniting all who use them, even if they all inflect their individual uses differently.

The ineffable, as it circulates in this company's rhetorical self-fashioning, cannot be defined ontologically, but it can be described operationally. Here, the ineffable, the transcendent, the spiritual are relational, both in their constitution and in their application. These metaphysical vocabularies are set in motion by shared ways of speaking and informal, everyday ritual performances: practices that affirm these mobilizing fictions as meaningful, even as they deny any ultimate meanings.

Spiritual mobilizing fictions in Hae Kyung Lee and Dancers point both towards and away from the subjectivities that constitute and deploy them. On one level, these fictions help stabilize individuals' self-definitions within the community. 'You are artists. Think like artists,' Hae repeats to her dancers. 'Make the commitment to think like artists. You have responsibilities to the Universe!' While the command to 'think like artists' is probably repeated countless times in countless studios, for Hae's company members and dancers from CSULA this formulation has special resonance.

California State University, Los Angeles, is an urban, commuter campus bordered by East Los Angeles; Monterey Park, an Asian suburb dubbed 'New China Town'; and Alhambra, a multi-ethnic, working-class community. The University enrolled 20,637 students in Fall, 2003; over 25 percent of the undergraduates are part-time (CSLA Facts #36). Most work full- or part-time in addition to taking classes. The average age of undergraduates is 25; many have families in addition to work and school responsibilities. Of the total student population, 84 percent are 'minority' (nonwhite) (CSLA Facts #36). Over 20 percent are classified 'immigrants,' 'non-resident aliens' or 'refugees.' Many CSULA students are the first in their families to go to college and pressure on them to choose 'marketable' degree programs is intense. Many opt for service-related majors; the top three undergraduate degrees granted are in Child Development, Criminal Justice, and Social Work, with Computer Information Systems and Accounting rounding out the top five.

CSULA students generally do not think of themselves as artists; family and economic pressures work explicitly to exclude such thinking. Sustained arts training in childhood is a luxury beyond the reach of many in the University's service area. A career in the arts, as my own CSULA students have often indicated, is seen as trivial at best and selfish at worst by families and communities who count on an individual's employability and earning power for survival. Jill Yip, a core member of Hae's company, was a Business major before meeting Hae and switching to Dance. Claudia Lopez, another core member, also started out as a Business major; on Hae's arrival at the campus, she was one of only a small handful of Dance majors.[6] Miguel Olvera, also a core member and Claudia's husband, was a Theatre major, though he planned on, in his words, 'doing the whole [entertainment] Industry thing.'

In this context, to think of oneself as an artist is to do much more than strike a blow for personal expression. It is a political choice with material consequences: relationships must be reconceived and managed; finances and security must be imagined in new ways, often without the benefit of family and community role models whose invocations might quell personal or parental anxieties or provide patronage; compensation for lack of, or inadequacies in, early training must be addressed.

Randy Martin observes, in a larger argument about university classes in modern dance and the instantiation of a mythology of an American self, that the 'guardians of technique mediate the rift between what local bodies produce and the general condition of a sustainable social order that those bodies may credit with the possibility of their own production' (1998: 154). Through participation in modern dance training, he argues, 'an imaginary affiliation with the nation-state's regime of authority may still get produced' (Martin, 1998: 154). Though Martin characterizes his reading of a university modern dance class as ethnographic, the demographics of the class, beyond gender, are not mentioned.

In the case of the CSULA students who went on to join Hae's company, however, the political trajectory of their choices and participation is very different. These dancers are, from the outset, inhabiting, not bridging, the rift between 'the local' and a larger state-sanctioned regime of self production. Modern dance technique widens this rift even further. Key members of the group, and the 'guardian of technique' herself, Hae, were not born in the United States. Their national 'origins,' their first languages, even where they go to visit relatives, are all outside, elsewhere, always framing the explicit constructedness, not seamless acceptance, of their identification with the state as 'American.' They are likewise subjects outside the typical trajectory of a dance career that begins with artistic self-fashioning in childhood, including rhetoric that moves unproblematically into fantasies of the artist as autonomous agent of her own design, or even as privileged dilettante. And they are outside family and cultural conventions of sustainability, indeed necessity, as they are not only, in parental narratives, 'forsaking' economic security but doing so in favor of a realm of aesthetics that demographically hardly recognizes them among either its participants or its audiences. Here, among these dancers, to 'make the commitment to think like artists' is to assert one's own creative prerogative with the awareness that such an assertion is never pure and never without material and relational consequences and costs for self and others. It is to reckon with the demographic fact that, as a modern dancer of color, the technique to which you commit could barely have imagined you.[7]

Spiritual, mobilizing fictions in Hae Kyung Lee and Dancers affirm self-definition, both by stabilizing the notion of artist as identity and by unsettling and relativizing the notion of 'self.' A dancer must think like an artist because s/he has 'responsibilities to the Universe.' If 'the Universe' is a transcendent abstraction, the responsibilities are not. Hae herself detailed the material labor mobilized by this metaphysical formulation in performance as part of the program 'The Horse's Mouth Greets the New Millennium' (Japan America Theater, Los Angeles, 22 July 2000). As the program explains: '*The Horse's Mouth* is a live, dance/theater documentary in which each participant tells a story from their life; then performs his/her own movement phrases – first in place, then traveling through space, sometimes improvising with another dancer... The participants are selected from the dance community in the city in which the piece is performed....' Hae began her story by recounting the conventions governing her early training in Korea; the teacher was an unquestioned authority and, in addition to taking class, students' responsibilities included labor in service to the community. Hae told the audience she had to 'clean the floor.' 'Now,' she said with no trace of resentment, after 13 years as founder and head of her own company, after a professorship and numerous national and international awards and fellowships, 'I am still cleaning the floor.'

She meant it literally. Hae and her students clean the floor of the campus performance spaces, not as some contrived exercise in enforced humility, but because the floors are dirty. Without this labor, they wouldn't get cleaned for rehearsals or late afternoon and evening classes. On performance days, generally weekends when no University janitorial staff is available, they clean the restrooms. Company members who, after securing their MAs in Dance, joined the CSULA part-time faculty, volunteer their time to stage manage Hae's students' shows and graduate thesis performances. Outside and at intermission they sell cookies and soda to raise money for scholarships to support current dance majors.

To think like an artist with responsibilities to the universe is, in this company, to serve – not some fantasy of one's own talent, but the community both narrowly defined as company members and broadly construed, including those still in training, audiences, and a world that could be made better through right action. This is daily service to other bodies and, as Hae says, 'to the work.' The referent of this 'work' is productively vague. It may involve learning choreography; it may include the labors of grieving or cleaning the floor. As Martin, Franko and others have indicated, modern dance technique is particularly amenable to rhetorics of autonomous agency, a kind of corporeal Manifest Destiny cast as distinctly American, over against the 'ossified' European tradition of ballet. Here, 'the work' is a metaphysical instability that frames dancers' relationships with each other and with their art in terms different than those of ownership and ambition. This ethic of service, cast in metaphysical terms, is also performatively constituted in a number of community practices and rituals.

'Giving energy'

Hae is describing the logistics of Ancient Mariners; she is very pleased with the Grand Performances tech staff. They seem to have joined the company community. This is particularly important because, while the performance is one night only, there are weeks of on-site rehearsals open to the public as they pass through the Water Court. 'Backstage' negotiations are all onstage here. Hae brings food to every rehearsal and has invited the crew to share these meals with the company, despite the fact that the producing venue is officially responsible for feeding its tech staff. It was so funny, she says. I was giving energy to the kids [her dancers]. They lined up. Then I looked and the crew lined up. They wanted me to give energy to them too.

Energy, ethics and corporeal commitments

As I have noted earlier, technique is, in one respect, a pre-existing conversation that expands to include us as we are birthed into dance; this conversation offers an almost utopian grammar for organizing bodies, history and desire. Yet, as it is inhabited, enacted, set to music, put into play, technique becomes, not only local, but an alibi or a shorthand for complex webs of languages and practices, both corporeal and metaphysical, that bind particular subjectivities to one another and, in so doing, authorize community. Many such praxical webs bind and authorize this community, but one in particular is an exemplary mobilizing fiction.

While 'energy' circulates as a kind of master trope through many dance contexts, in Hae's company the term signifies more than intensity, momentum or heightened affect. Here, 'energy' is a technology of group cohesion and of intrapersonal focus, spanning both and linking them together in community. Like all metaphysical aporias, 'energy' as used here is both very empty and very full of communicative potential. Mandy Gamble notes in her thesis on the group that it is something every company member understands, but cannot necessarily define (1997: 61).

'Energy' is a metaphysical performative, not an ontology; it finds expression as enactment. One doesn't have to believe anything about it in order to participate or, for that matter, to not participate. Hae 'gives energy' to the dancers through touch and meditative practice. Dancers work on their energy through right thinking and right living in and beyond the studio. There are no specific ritual protocols organizing the giving of energy; varying degrees of formality are involved. Incense may be burned, a dancer may simply be hugged, or Hae may say she is sending an individual energy.

As all involved are quick to point out, 'energy' is not religious; it is not doctrinal. Though Hae is an Asian Buddhist, energy is not characterized as *chi*, nor is the ritual vocabulary involved in giving energy 'denominational.' It is difficult for dancers to articulate what actually happens when they 'get energy,' though it is very clear to them that something does happen. While 'energy' is an insider discourse, its resistance to denotation places participants always already consciously outside it. Consider 'energy' a productive instability deployed as a mobilizing fiction, one that reinforces individual integrity and group solidarity without discursively or 'theologically' fixing either.

In *Witnessing: Beyond Recognition*, Kelly Oliver argues for a non-Hegelian approach to subjectivity and an ethics that dispenses with the view of relations with others as struggles for recognition. Energy is central to such an ethic. Oliver explains: 'All human relationships are the result of the flow and circulation of energy – thermal energy, chemical energy, electrical energy, social energy. Social energy includes affective energy, which can move between people. In our relationships, we constantly negotiate affective energy transfers. Just as we can train ourselves to be more attuned to photic, mechanical, or chemical energy in our environment, so too can we train ourselves to be more attuned to affective energy in our relationships' (2001: 14).

Whatever else energy is for Hae and her dancers, it is certainly social and affective. Heightened awareness of this relational energy between dancers is not only useful for company harmony and cohesion. Physical safety depends on it. Hae's choreography is demanding; there are a variety of challenging lifts and spins that, even while appearing playful, are taxing and dangerous if a dancer is not aware of his/her body and others in space (Figures 4.1 and 4.2). Dancers are twirled close to the ground; in pieces like 'Catch My Drift,' they jump and roll over one another so quickly that they must initiate a move by anticipating, not actually seeing, where another dancer will be. The demands of this choreography dictate the kind of relaxed, but focused attention 'energy' marshals and provides; more than this, the work requires a strong sense of responsibility, both for one's own body and those of others with whom one shares the space. Oliver observes: 'Because our dependence on the energy in our environments brings with it ethical obligations, insofar as we *are* by virtue of the environment and by virtue of relationships with other people, we have ethical requirements rooted in the very possibility of subjectivity itself. We are obligated to respond to our

Figure 4.1 Miguel Olvera and Claudia Lopez of Hae Kyung Lee and Dancers. 'Blank Slate' from *Seventh Heaven* series. Aratani/Japan America Theatre, 2005. Los Angeles, California, USA. Choreographer. Hae Kyung Lee. Photographer: Chin Fong

environment and other people in ways that open up rather than close off the possibility of response. This obligation is an obligation to life itself' (2001: 15).

Thus, energy does more than crystallize and encourage individual focus, more than performatively promote group solidarity, though these are vitally important. Energy reinforces an ethics of obligation, of service onstage and off, an ethics of presence to others as bodies rooted in these dancers' physical interdependence. Despite what the metaphysical vocabularies used to communicate it might suggest, this ethics of service is non-ontological. It is, however, corporeal, rooted in physical proximity, in touch. As John Durham Peters asserts:

Touch, being the most archaic of all our senses and perhaps the hardest to fake, means that all things being equal, people who care for each other will seek each other's presence. The quest for presence might not give better access to the other's soul, *per se*, but it does to their body. And the bodies of friends and kin matter deeply . . . . With his war on 'the metaphysics of presence,' Derrida is right to combat the philosophical principle that behind every word is a voice and

Figure 4.2 Kishisa Ross and Julio Moran of Hae Kyung Lee and Dancers. 'Blank Slate' from *Seventh Heaven* series. Aratani/Japan America Theatre, 2005. Los Angeles, California, USA. Choreographer: Hae Kyung Lee. Photographer: Chin Fong

> behind every voice is an intending soul that gives it meaning. But to think of the longing for the presence of other people as a kind of metaphysical mistake is nuts. (1999: 270)

Daily rituals like those involving energy shore up an ethics of service and presence within the community of dancers. These rituals make explicit the social responsibilities each dancer has for the well being of others that includes, and goes beyond, the responsibilities for enacting demanding choreography safely with focused attention and precision. Further, these social, ritual performances also make visible an ethics of authority based on service. Energy is framed as a gift; Hae 'gives' energy and her dancers receive it. But this authority in/as service is not only metaphysical. Hae feeds her dancers, bringing food to every rehearsal, every performance. She buys glasses and medicine, and pays for visits to the acupuncturist. She has taken core members of the company to New York, to Paris, Switzerland, Germany, London and Korea, paying for plane fare, lodging and dance classes. She has secured employment for company members, bought them clothes and tickets to performances, all on a professor's salary. 'They are with me,' she says, 'so they are my responsibility.'

'Responsibility' here, as in the construction of an ethics of service for the community of dancers, is not a burden. It is an ethics of care that supports the other's response: response-ability. As Oliver explains, in Levinasian terms: 'I am responsible for the other, for the other's response, and the other's ability to respond' (2001: 206). Hae knows very well the kinds of material sacrifices her dancers make on a daily basis to think of themselves as artists. If they are to be response-able to 'the work,' she reasons, she must do everything she can to make such responses possible. This ethic is grounded in metaphysical rhetoric that eliminates more conventional issues of authority; Hae does not support her dancers because she feels they will then 'owe' her. She makes this clear using spiritual formulations which affirm her authority in/as service, defer the ultimate source of that authority, and diffuse any expectations of immediate *quid pro quos* from company members. 'My spirit tells me,' she says, 'that I must take care of them and my spirit helps me do this. I don't think about the money. Somehow it will be there. And someday they will do this for other students.'

Like all ethical relations involving authority, the other may opt out, may choose not to respond. Core members who joined Claudia, Miguel and Jill during Hae's first years at CSULA have indeed opted out over time. In some cases, dancers found that they did not comfortably fit the community's aesthetic ideals or its mobilizing fictions. In other cases, outside pressures to live more conventionally secure lives became too difficult to resist. Still others departed for reasons unknown and unsaid. Key departures are wrenching for all involved. The cohesiveness of the community and the intense time commitments to practice and performance mean that those who leave lose what is, for most, the primary subject position they've inhabited for years, forged through daily routines and close relationships. Dancers who remain must deal with losing a partner, or even a roommate; corporeal commitments must be reconfigured in choreography and in life. While Hae herself is deeply affected by these departures, she deploys energy as a mobilizing fiction to manage transition and loss. 'Don't hold on,' she admonishes herself and her dancers. 'Let go. Everyone has their path. To stay stuck is no good.' On one level, this is an aesthetic pronouncement; it is visible in the liquid quality of movement in Hae's technique and choreography, about which more below. But more than this, the admonition to let go, to embrace social and affective energy as an ethical process rooted in corporeal presence, is a way of acknowledging that, when dealing with bodies in time, onstage and off, stasis is not possible.

'Human on stage'

I am having lunch with Miguel Olvera and Claudia Lopez. We are talking about a recent performance by The Trisha Brown Dance Company. Though the dancing was interesting and enjoyable, elements of the choreography were disappointing. With one exception, the company appeared to be all Anglo; dancers wore gender appropriate costumes and men lifted women. This is so different from performances by Hae's company, where Anglo dancers are the distinct minority – one or two members at most. Women lift men as a matter of course, and size is no object to the extent that when a new dancer, a vegetarian and, in Claudia's words, 'tiny, tiny,' joined the company, Hae told her to eat more protein or she would never be strong enough. Moreover, men dance in women's clothes: in evening gowns, halter tops and floral raincoats, with hair in pigtails or buns or braids. 'Can we talk about the play of gender and ethnicity in Hae's work?' I ask. 'Oh, we don't think in those terms,' Claudia says firmly. Miguel adds, 'We are human on stage.'

Metaphysical performatives and alterity

> *How substantial is the difference between the other as existing entity, such as a human being, to which I am enjoined to respond, and the other as that which beckons or commands from my mental sphere as I engage in a creative act?*
> (David Attridge, 'Innovation, Literature, Ethics', 1999)

Claudia and Miguel, and other members of Hae's company, are neither naive nor politically unaware. Quite the contrary; they have not had those luxuries. Miguel, for example, was a member of the highly political Latino company *La Teatro* after receiving his undergraduate degree. In a narrative he gave to Mandy Gamble as part of her thesis on the company, he recounted an experience with racial profiling: being singled out and followed by security when he stopped at a drugstore after rehearsal:

> I decided to turn the tables and not be the victim. I decided to invade his space. So I pretended I was still searching for whatever. And I

moved really close to him as he pretended to be looking at something else; I was so close to him I bumped his shoulder. He began to move away, and I slowly followed him as if I was searching for something. I kept close to him for awhile, then I went about my business. He disappeared.

...I realized that I [was] perceived as a 'Chicano,' 'Hispanic,' 'Latino,' Brown man, with shitty clothes; I became a threat to their establishment. (1997: 40)

In her narrative describing an encounter with the predatory male gaze, Claudia described her desperate search for a job after leaving her parents' home prior to joining the company. Answering an ad for 'photographic work,' she found herself in the office of an agent who asked her to take her top off (Gamble, 1997: 38–9). Jill Yip's account merged gender and ethnicity; she related how, on a lunch visit to Chinatown with three male friends, she was not given a menu and was served last (Gamble, 1997: 36–7). The relative banality of the accounts testifies to their ubiquitousness; whatever these dancers mean by 'we don't think in those terms,' it is certainly not 'race, gender and class don't matter.'

In the first half of the twentieth century, modern dance technique had both an entrepreneurial and an activist orientation to alterity. Sometimes technique emphasized the extremity and visibility of difference, which it could then uphold as exotic wonder. Sometimes technique mobilized difference as a reaction against oppression; sometimes difference was recoded as raw material for an autonomous genius. Ruth St Denis is perhaps most exemplary of the entrepreneurial turn, invoking Asian movement vocabularies in service of what Sally Banes describes as 'a quintessentially American quality': 'Although Asia was old and America was new, the two continents seemed to share something – an imagined simplicity and spiritual wholeness – that Europe lacked' (1998: 91; see also Desmond, 2001). In a much more radical vein, Ellen Graff (1997) details how working-class and Jewish bodies were organized by modern technique, and vice versa, for political ends in New York from 1928–42.

In this company, subsuming difference in the rhetoric of 'universal humanity' as metaphysical mobilizing fiction has multiple uses for Hae and her dancers. First, it is a way of claiming the privileges and prerogatives of the 'artist subject,' whose white masculinity usually goes unmarked. Second, and closely related, is resistance to what Guillermo Gomez Peña has called culti-multuralism: fetishizing difference to add texture to arts communities, generating both tokenism and, ultimately, in-difference. Hae has expressed repeated frustration at expectations that she 'should dance only "Korean" dance, whatever they think that is – pretty Asian girl with a smiling face and flute music.' Multiculturalism, particularly in Los Angeles in the mid-1980s to mid-1990s, was for many an art ghetto.[8] Finally, because company members occupy multiple axes of difference, they have multiple relations to visibility, most of them burdensome: hypervisible Latino, hypersexual Latina, invisible Asian woman. Through the metaphysics of 'universal humanity onstage,' they can rewrite relations to visibility in which the particularity of their bodies is always exposed, but burdensome racist stereotypes that accrue to being (un)seen can, at least within the choreography, be suspended in favor of other possibilities. Consider one of Hae's seemingly most playful pieces, *No Room to Move* (1995), performed at the Getty Museum (21 April 2000). The dancers, in evening gowns (including the men), sat staring at the audience for what seemed like an oddly long time. The costumes, complete with cat's-eye glasses, and the faces the dancers made as they stared at the predominantly Anglo viewers (grimaces, squints, etc.) elicited some uneasy laughter. Perhaps this laughter arose from the incongruity of the scene; perhaps this was anticipatory chuckling, filling the space until something happened. Or perhaps this was nervous laughter for another reason. As Richard Terdiman puts it: 'Does not the other still have a gaze which returns our own? Does the other's gaze not put pressure on our own?' (1991: 8). The very particularity of these bodies, costumed so that all of them appear to be in drag, yet still clearly artists of color, suggests that this pressure of a returned gaze might account for some of the audience's uneasy giggles.

Ethics is predicated on alterity, as is metaphysics. 'The One' polices right action *vis à vis* an other. In the metaphysical mobilizing fictions that organize this company, there is no legitimating, specified, essential One, but alterity and difference are ethically organized all the same. Central to this operation is Hae's view of the relationship between art and life. In contrast to artists like Martha Graham, who turned to dance to give voice to the 'genuine' that might be masked in life or, on the other hand, members of the Judson Dance Theatre who brought pedestrian movement to the stage to reinvigorate dance, Hae does not conceive of art and life in binary terms. She frequently admonishes her dancers during rehearsals:

> Dance is not only dance. Dance is life. How you live is how you dance. You can see it in the work, the energy of how you live. Doesn't matter who you are. If you learn this you can do so much with your

life. Art and life are not two. Don't make that mistake. Art and life are the same thing. [If y]ou lie and cheat in your life, you are not an artist, doesn't matter who you are, how great everyone says you are. You must be so clear, so honest, all the time. That's why I tell you, 'Don't lie to me, honey. I will always know because I can see and others will see. And you know too.' Doesn't fool anybody. If I am not honest, the Universe – I could not do what I do. I could not make these pieces.[9]

The act of creation itself is cast as a consequence of responsible living. Further, this ethically-informed view of creation emphasizes response-ability to an other, here the 'universe,' rather than autonomous agency, which then gets transmuted into right action towards others rather than creative genius. Further, a dancer's responsibility to others translates into a gift of realizing an other: the creative product.

Rhetorically constructing the work of art as a metaphysical product of social response-ability is an inventive and useful way of insuring group trust and cohesion around a central mission, albeit one that each dancer understands differently. Yet the view of creation as alterity is not unique to those who deploy the spiritual to characterize it. Derek Attridge offers an account of this 'otherness' in writing, though the creative process he describes is analogous to that of the universe enabling choreography:

In a curious way, the ideas I have not yet been able to formulate seem to be 'out there' rather than simply nonexistent. My experience has an element of passivity, of attempting to heighten my responses to hints of relation, to incipient arguments, to images swimming on the edge of consciousness, an element of letting them come as much as seeking them out. When I write a sentence that seems just right... I am not able to say how it came into being, but I can say I did not produce it solely by means of an active shaping of existing, conscious, mental materials.   (1999: 21)

In Hae's account of the creative process, 'hints of relations' with the ineffable are linked to the materiality of relations with the social.

Here too, metaphysical performatives are rhetorically powerful tools of community building; they bind alterity in the company to alterity in creation, embedding both in the imperative to be socially and corporeally present to other bodies and to 'the work' they produce. Yet such performatives are also useful for exploring and connecting seemingly incommensurable aspects of the creative process itself. Invoking 'the universe' links the efforts of choreography and rehearsal, those products of sweat and sacrifice, to enchantment, the senses of inspiration and grace when artists bring forth something they didn't know was in them. Such metaphysical performatives create a discursive zone marked by the simultaneity of seemingly incompatible vocabularies marshaled on behalf of explaining the creative process. These include love and labor, discipline and freedom, self-awareness and self-abandon. A vocabulary rooted in behavior or cognition deals in commensurabilities: cause is always commensurate with effect, input with output. In Wai Chee Dimock's words, this kind of language 'prohibits irregularities and it also ignores miracles, occurrences so extraordinary as to exceed its grammatical description. It is thus a language of the lowest common denominator, one that, if adopted, would explain why we might have no quarrel with the world. But it would not explain why we might love the world' (1996: 110).[10] A metaphysical vocabulary, particularly one as porous as that undergirding this company, acknowledges that there is always something lost in attempts to translate accounts of the creative process into straightforward denotative, or even connotative, language. Yet simultaneously, something is gained in the ability to tactically unify contradictory aspects of that process in a shorthand meaningful to community members.

For both Hae and Attridge, alterity becomes the language borrowed to translate this slippery, contradictory supplement, or antidote, to an authorial and social agency that emphasizes *self*-importance. Attridge continues:

> Hence the appropriateness of the alternative reading of [the] genitive construction in *the creation of the other*, according to which my text, and perhaps something of myself, is created by the other. This should not be taken to imply a mystical belief in an exterior agent; rather, it indicates that the relation of the created work to conscious acts of creation is not entirely one of effect to cause. The coming into being of the wholly new requires some relinquishment of intellectual control and *the other* is a possible name for that to which this control is ceded.
>
> (1999: 20; emphasis in original)

While they may or may not have mystical beliefs in exterior agents, Hae's dancers explicitly address the alterity of/in creation. As Miguel observed, 'When [Hae] tells us what she wants us to do for a piece, sometimes I think there's just no way. But then we work and bring out something we never knew we had in us.' Claudia added, 'Hae is never worried. She says, "Relax. Don't think about it. It will be there." And it is. Who knows where it comes from.'

Here the work, 'it,' is an other, a stranger; 'Who knows where it comes from.' Further, this other of the work makes the body executing it a stranger too: 'We never knew we had [it] in us.' Both of these 'strangers' demand respect and call for response-ability, for dancers 'to allow themselves to become engaged even to the point of being in a sense remade' (Buell, 1999:12). 'Remaking' is not 'self-authorship,' not the entrepreneurial relation to agency so characteristic of the rhetoric of early American modern dance. Hae continually defers her own agency as a choreographer to the agency of ethical living ('If I am not honest . . . I could not make these pieces.'), even as she defers her agency in her dancers' material support to spiritual obligation. Moreover, she encourages her dancers to do the same. Indeed, core members of the group all seem to 'speak away' their authority as dancers in a manner Sells calls 'apophatic': 'Apophasis can mean 'negation,' but its etymology suggests a meaning that more precisely characterizes the discourse in question: *apo phasis* (un-saying or speaking-away) . . . . Any saying . . . demands a corrective proposition, an unsaying' (1994: 3)

When I listened to Hae describe how she generates her dances, or when I asked her dancers to tell me how they came to know what they never knew they had in them in rehearsal and in performance, the responses are replete with the 'paradoxes, aporias, and coincidences of opposites' Sells sees as characteristic of apophasis (1994: 3):

Miguel: I – you – we just get it. Or we do it, and then recognize we got it. Or we just know it all along but not clearly. It's hard to explain.

Hae: This is a gift, a blessing from the universe. You earn this by being a good person in the world. It is your mission to make the world better with your work and your life.

As Sells observes, 'Real contradictions occur when language engages the ineffable transcendent but these contradictions are not illogical' (1994: 3). Thus, dance is, simultaneously, a gift, a blessing, a kind of reward and a mission. It is known imperfectly all along, realized retrospectively, and the subject of present-moment epiphany. It's hard to explain but not hard to enact. Two interrelated aspects of the ineffable hold these apophatic formulations together. They are 'the universe' and 'the world,' the generative link between the transcendent and the community implicit in Miguel's 'it' and his final discursive choice of 'we' in 'I – you – we.'

Here, the ineffable other of creation makes demands on material others, both in the company and beyond. Responsibility to one's art involves abandoning a measure of control over the product in favor of a deliberate process of ethical living and service that remakes dancers as artists in solidarity with those they serve. This is not a solidarity of racial, gender or class amnesia. It is, instead, a kind of utopia organized and maintained by metaphysical mobilizing fictions, then incarnated in technique and shared with the world. As Jill Dolan observes, idealism, like that undergirding the ethics of service and the spiritual vocabularies invoked by this company, funds a politicized view of utopia. She writes, 'I find this notion very rich, the idea that, in order to pretend to enact an ideal future, a culture has to move farther and farther away from the real into a kind of performative, in which the utterance, in this case, doesn't necessarily make it so but inspires perhaps other more local "doings" that sketch out the potential in those feignings. . . . The utopia for which I yearn takes place now in the interstices of present interactions, in glancing moments of possibly better ways to be together as human beings' (2001b: 457). Yet, in Hae Kyung Lee and Dancers, metaphysical performatives make, not fake, generative sociality and solidarity, even if the mechanics of this making are understood differently by all involved. Further, there is no movement away from the real; that would be an evasion of the responsibility corporeal and social presence demands in choreography and in life. As Hae has said on many occasions: 'Maybe you don't understand me. Maybe you go home and think this is crazy. That's okay. Better you say to me, "Hae, this is crazy shit" than you sit here and smile and lie to my face, "Oh, yes, Hae." But [it] doesn't really matter either way. Do what the universe tells you. Live a good life to be an artist.' Here, metaphysical performatives move to organize a utopia of ethical relations in art and life in the here and now.

Metaphysical vocabularies and reading choreography

> *I can easily believe that there are more invisible than visible beings in the universe. But of their families, degrees, connections, distinctions and functions, who shall tell us? How do they act? Where are they found? About such matters the human mind has always circled without attaining knowledge. Yet I do not doubt that sometimes it is well for the soul to contemplate as in a picture the image of a larger and better world, lest the mind, habituated to the small concerns of*

daily life, limit itself too much and sink entirely into trivial thinking.
But meanwhile we must be on watch for the truth, avoiding extremes,
so that we may distinguish certain from uncertain, day from night.
(T. Burnet. Epigraph to Coleridge, 'The Rime of the Ancient
Mariner', 567; italics in original)

A metaphysical vocabulary serves dancing communities in yet another way. This vocabulary, with its productive porousness, its invitation to reflect on 'the challenges and rewards of converting immaterial to material and vice versa' (Kuspit, 1986: 324), offers a representational frame to organize, 'as in a picture,' the antinomies and ambiguities of, and within, choreography. Technique generally binds bodies together in a common corporeal idiom. Metaphysics serves as a particular kind of supplement, one that gives this idiom an affective, rhetorical half-life beyond the denotative and the corporeal. It mobilizes performers by offering images that contain or resolve seemingly contradictory physical operations, subsuming them in a 'higher' or 'larger' unity or logic. To return to Burnet's final sentence in the epigraph above, the 'invisible beings' of metaphysics invoked by Hae Kyung Lee and her dancers ('the universe,' 'energy,' 'spirit') do not so much help the company 'distinguish' between 'extremes' as hold these in tension, making physical and discursive paradoxes communicable and intelligible.

Much of Hae's choreography presents, but does not resolve, productive paradoxes, including physical and thematic ones. 'Beyond the Ancient Mist,' part of her *Silent Flight* series, is especially illustrative. The dancers begin clustered together in a corner upstage; dressed in flowing white gowns, they each hold long fabric panels, also white, that they seem to pour between outstretched and upraised arms (Figure 4.3). They move across the stage on a diagonal with rolling steps, both forward and back. When coupled with the gentle undulations of the dancers' torsos, the overall sense is of waves 'closing and closing in, but never quite,' as Elizabeth Bishop's 'Crusoe' might say (1984: 162). The faces of the dancers are both focused and completely serene. Yet the fabric pouring between their arms seems, as they move across the stage, to take on an energy belied by the leisurely lyricism of their movements and expressions. While they continue to progress using their gentle, rolling steps, the fabric flies higher and higher into the air; it becomes animated, crackling as it rises and falls. The dancers appear to be using no more visible force or momentum than before; the fabric seems to have taken flight of its own accord. The overall effect of this movement vocabulary seems, technically, both timeless and utterly contemporary

Figure 4.3 Claudia Lopez, Claudia Medina, and Patricia Warren of Hae Kyung Lee and Dancers. *Beyond the Ancient Mist.* California State University, Los Angeles State Playhouse, 1999. Los Angeles, California, USA. Choreographer: Hae Kyung Lee. Photographer: Roger Burns

and, thematically, like a tension between the gentle waves of history and the crackling flights of the present becoming past. These paradoxical elements are further echoed in Hae's program note accompanying one piece in the series: '. . . *I am the stones my ancestors walked on. Floating like paper, like light. . .* ' (program for DanceWest, The Western United States Platform of the Seventh Recontres Chorégraphiques Internationales de Seine-Saint-Denis, Luckman Theatre, Los Angeles, 14 January 2000).

The inspirations for this movement sequence, Hae tells me, are traditional Korean dances, particularly shamanic ones. Yet, for all of the serene upward and forward trajectory of this passage, *Silent Flight* does not simply reproduce the lightness and vocabulary of classical Korean dance. (As *Los Angeles Times* dance critic Jennifer Fisher so aptly stated in metaphysical, if tongue-in-cheek, terms: 'Nobody uses the God-given ability to float quite like Korean dancers' (2003: E4)) These dancers' advance across the floor blends rising and sinking with a highly, though subtly expressive upper body, controlled arms and faces, and the drama of the billowing fabric. The tensions between these various components are highly visible, but not simply blended or resolved. Further, this piece suggests, in a larger way, Hae's unique alchemy of modern, postmodern and classical Korean elements by exploring and preserving some particularities of each, as well as their points of convergence.

'The hardest thing for them are the faces,' Hae has observed, referring to the serenity of the dancers' expressions. 'It is so hard for young people to really get this. The shaman is older and with age comes that wisdom. So I try to tell them [her dancers], "Young body, old spirit. Concentrate but go above. Be here and above. Be light but be into the ground." First they look at me like, "Hae is crazy." I have to tell them, finally, "Okay. Don't try to explain it back or think about it in words. Trust me and trust yourself and just do it." Then they just get it, finally.'

For Richard Schechner and Victor Turner, the performer, in ritual or on stage, exists in between possibilities that include self-consciousness and self-erasure. Schechner draws on psychoanalyst D. W. Winnicott's notion of the transitional object to characterize this 'in-betweenness' as the 'not-me and the not-not-me' (1985: 110–12). For Hae and her dancers, the process is different. Dancers must be *in*, not *between*, 'here and above.' They are directed, not to some transitional space but to two distinct states ('young body, old spirit') they must occupy simultaneously.

Hae's seemingly contradictory choreographic and technical imperatives ('be light but be into the ground') generate equally contradictory receptions from her dancers: from bald dismissals to eventual understanding. I would suggest that what leads her dancers to 'get it, finally' is their familiarity with the metaphysical formulations that organize their daily sociality as a company. Specifically, metaphysically-based paradoxes in linguistic idioms are ultimately intelligible, if not automatically or simply meaningful, because dancers are immersed in the productive porousness of metaphysics as a social idiom.

In *The Mystic Fable*, de Certeau explores the work of metaphysical language in context as 'manners of speaking': 'From this point of view also, as J. Baruzi wrote, "the mystic language emanates less from new vocabulary than from transmutations performed within the vocabulary borrowed from standard language." Properly speaking, it is not a new or an artificial language. It is the effect of an elaboration upon existent language, a labor applied to the "vulgar" tongues . . . , but extending also to technical languages. The uses that define it reflect the operations carried out by speakers' (1992b: 142–3). Among these operations and transmutations, oxymoron ('be light but be into the ground') has a special place: 'It makes a hole in language. It roughs out a space for the unsayable. It is language directed toward nonlanguage,' as in 'Don't try to explain it back or think about it in words' (1992: 143). Through her use of oxymorons, Hae points to the nonlinguistic space of 'trust' between dancer and choreographer, dancer and dancer, dancer and self. This is a nonlinguistic space shored up by rituals of energy, a trust buttressed by the ethic of service to 'the universe' and to 'the work' that all share. In this transformative, superceding space organized, not by language, but by physical and social presence, the company can engage and produce possibilities seemingly beyond those that can be denotatively stated. They can implement the 'unsayable' because the larger unities of which they are a part, art and life, right living and successful dancing, self and other, bridge the 'holes' in both language and the technique that language vainly, and vaguely, attempts to describe.

Metaphysical manners of speaking, and particularly oxymorons, serve more than a pedagogical function in Hae's work with her dancers. They also characterize thematic elements in her dances and her larger structural, compositional aesthetic. Thematically, Hae's choreography explores the terrain that both links and separates the two poles of her company's metaphysical ethos: 'the universe' and 'the world' or, in Homi Bhabha's terms, 'aura' and 'agora.' Bhabha asks: Is there a passage from the *aura* of rapture, of moving, seemingly outside and beyond the worlds, to the *agora*, or marketplace, of negotiation with the world? (1996: 10). The answer, he suggests, involves less a passage from one state to another than a zone marked by the simultaneity of both. He continues:

> For art has the capacity to reveal the almost impossible, attenuated limit where aura and agora overlap, to find a language for the high horizons of humanity itself and – in its finest selves, its inspired othernesses, its visionary styles, its vocabularies of vicissitudes – to reveal its own fabulation, its fragility, at the moment of its articulation. It is not that art's auratic ambition is somehow humbled by the unspectacular wisdom of everyday life – nothing so simple. What is revealed between the ecstatic and the everyday is a *mediatory in-betweenness* that belongs neither to aura nor agora – and this, in all its mystery and its ordinariness, is 'the human position'.   (1996: 10)

Hae would add, from the perspective of the metaphysics of corporeal presence undergirding her company, that this 'mediatory in-betweeness' of mystery and banality characterizing 'the human position' is *both* aura *and* agora, as much as 'neither-nor.'

Her recent work, *Cross the Wounded Galaxies* (12 October 2002, Japan America Theatre, Los Angeles), is an especially representative example of the 'almost impossible, attenuated limit,' where 'universe' and 'world,' aura and agora cohere in the thematic elements of her work. All three movements of this piece hold solidity and ethereality, deliberation and abandon, pragmatics and grace in unresolved tension by exposing the profound, corporeal humanness of each term. The second and third movements of this dance are particularly illustrative. In Part 2, the dancers move, as couples, diagonally across the floor, each *pas de deux* a kind of lyrical, slow-motion acrobatics. Jose Reynoso, a new company member, lifts Kishisa Ross, sets her down, rolls to the floor, balances her on his shins, supports her with his abdomen. Claudia Lopez carries Miguel Olvera behind her back, head down, before the two of them melt into poses where he, too, is held in the air on the soles of his wife's feet, or balanced on her shins. Three couples make this slow, diagonal passage across the stage; every hand placement is visible, every alignment precisely and meticulously executed. Yet despite the utter visibility of the corporeal mechanics of this passage, there is something more, some surplus that exceeds the careful, indeed painstaking pragmatics of the *pas de deux* themselves. It is as if one is watching intimacy being carefully constructed before one's eyes even as one is utterly at a loss to account for its gentle and elegant intensity. Perhaps it is because the couples' bodies are always in contact through every transition that the details of their corporeal relationships with each other seem, for all their visibility, unable to fully account for the intimate mystery of the contact itself.

Movement 3 offers a completely different set of images that, nevertheless, merge the 'ecstatic' with the 'everyday.' If Movement 2 centered on the excess and precision of intimacy, this section explores the corporeal labors of transcendence. Here, three sets of paired fabric panels flow from ceiling to floor of the proscenium stage. Miguel Olvera, Jose Reynoso and Julio Moran, who joined the company for this performance after taking Hae's classes at CSULA,[11] manipulate the fabric, move around and through it and, finally, wrap themselves in it and fly. The result is extraordinary, with these flights filling both the entire stage space and, as Miguel Olvera swings more and more widely, part of the audience area as well. Each dancer hangs cantilevered across the fabric as if it was a trapeze. They hang from the fabric head down; they twist it together, jump on, and spin as it unwinds. But this is no simple virtuosity. While their faces remain serene, their muscles strain; their own care with the precariousness of their positions is clear. They are beautiful and daring, but not careless or heedless. They negotiate the fabric rather than mindlessly master it. In the final sequence, each inhales and, visibly summoning up momentum, launches into a vigorous climb up the fabric. As the music ends, they are still climbing: no simple transcendence gained, but instead the labor of spiritual ascent with, as Bhabha would say, 'all its mystery and its ordinariness' in process.

The labor and effortlessness, the banal responsibilities to 'the work' and the world ('cleaning the floor') and the metaphysical potential of the artist serving 'the universe,' have their echoes in Hae's compositional aesthetic. Here, too, she uses the oxymoron as a metaphysical 'manner of speaking' to point to a space, and a logic beyond that of conventional communication. Specifically, she often plays with narrative form in her longer pieces, even as she disrupts simple story continuity. *Ancient Mariners* (2000), the first dance mentioned in this chapter, is an especially rich example. Interestingly, Hae was completely unaware of Coleridge's poem when she choreographed and titled this work. Nevertheless, the vision of a performance, impelled by 'invisible beings,' that both atones for error and points toward grace resonates strongly across both the dance and the poem. Indeed, this notion of atonement and grace, linked to a larger metaphysical logic, figured strongly in Hae's characterization of *Ancient Mariners* as 'the universe's piece.'[12]

Ancient Mariners deploys a standard dance 'grammar.' It moves from an ensemble introduction through solos, trios and *pas de deux* which, in more conventional choreography, play out content and/or formal themes. The piece ends with the apparent conclusive force of the *corps*. Embedded within this grammar are props that function as invitations to read for narrative sense: a boat, a torch, fish. These visual cues seem to ask to be reassembled into one coherent story or, at the least, to be read as fragments of multiple stories.

Several factors simultaneously work against and complicate this interpretive invitation. First, themes cannot be smoothly mapped onto the grammatical shifts from one combination of dancers to another. Consider what is, for me, the most extraordinarily beautiful moment in an extraordinarily beautiful dance: the solo/*pas de deux*/trio featuring Jill Yip, Claudia Lopez and Miguel Olvera. Note that the formulation 'solo/*pas de deux*/trio' itself indicates the slippage between formal configurations of dancers in the piece, one mirrored by its play of themes.

The sequence begins with Miguel Olvera moving across one level of the fountain/stage holding a lighted torch and pulling a small open boat, inscribed 'Ancient Mariners,' behind him. Claudia Lopez guides the boat from the rear as Jill Yip stands inside. Jill uses the boat not, as expected, to mark a place of solidity apart from the rushing water in the fountain that surrounds her, but instead as a vessel for acrobatic yet liquid movements that both echo and reframe the configurations of the water itself. Her legs, arms, torso, fingers seem to blur together in waves of movement, the flow punctuated by headstands, poses, cartwheels held with legs in the air as the fluidity of her dancing is punctuated by the briefest moments of stasis. Then Jill seems to disappear, or perhaps to dissolve, and Miguel and Claudia explore this liquid acrobatic vocabulary even more completely (Figure 4.4). Each lifts the other with remarkable ease; at moments it is impossible to tell who will lift whom, so seamless is the flow of the movement between them as, for example, when Claudia effortlessly lifts Miguel and transforms him into an oar to steer toward – where? Then, setting him down, she is herself lifted as if this passage were a kind of movement stream in which one dancer surfaces and submerges, followed by the other. Miguel and Claudia are soon joined by Jill and the three continue to embody the contrasts between stasis

Figure 4.4   Claudia Lopez and Miguel Olvera. Duet from *Ancient Mariners*. Grand Performances, California Plaza, 2000. Los Angeles, California, USA. Choreographer: Hae Kyung Lee. Photographer: Roger Burns

and flow, between mastering the water in which they are immersed and becoming it. Thus, the solo and *pas de deux* do not stabilize the centrality of the individual or the couple within the frame, as in conventional dance grammars. Instead, these configurations dissolve one into the other, highlighting the contingency of relations within the larger flow that contains them. Further, the dancers' undulating movements, punctuated by athletic stasis like a swimmer's stroke punctuated by breath, cross all these configurations, marking their thematic coherence rather than, as might be expected, discrete stages in *Ancient Mariners'* 'narrative' development.

In a related vein, Steve Moshier's score, performed live by the Liquid Skin Ensemble, moves forward through lyrical, hypnotic repetition with very subtle variations: no builds, no narrative cues. Indeed, this music is nonnarrative in the extreme. Like the dance itself, which attenuates rather than demarcates grammatical shifts, there is a sense of 'and then, and then,' but no simple 'and then, because,' despite repeated formal and content suggestions in the dance that 'because,' that narrative chronology and causality, was possible. Timothy Bahti reminds us that the lyric shapes and reshapes itself between pure circularity and pure linearity (1996: 254); it both moves forward and circles back upon itself. Certainly this is an exemplary description of the *Ancient Mariners'* score. But when this oscillation between circularity and linearity, repetition and progression in the score is complicated by hints at narrative and the smooth ambiguity of abstract movement in the vocabularies used by the dancers, the larger development of this dance over time becomes doubly complex. Poised between narrative and lyric invitations, each complicated in themselves, spectators must read between and through different logics of progression and development, not for the reassurance that some final resolution will knit these complexities together into meaning there for the taking – no such resolution is offered. Rather, the power of this work is precisely its oxymoronic disturbance of the dance lexicon, its opening of a space in which to imagine an experience of shared time in performance as itself unsayable, mystical, yet profoundly communal. To put it another way, the complex strategies of progression in music and movement that characterize *Ancient Mariners* do not compel 'through what they directly signify (this is precisely what they [refuse])' (Ahearne, 1999: 109). Rather, in the words of Jeremy Ahearne, they produce in the spectator 'a form of passage through and beyond conventional linguistic categories' for organizing and explaining how dance, or, perhaps, atonement and grace, move through time (109). The spectator is invited instead to carve out a fantasy space where both the specificity and the indeterminacy of this particular experience can persist unresolved, a metaphysical space mirroring the sociality from which 'the universe's piece' was created and performed.

Hae's dances are luminously beautiful though, as noted above with *No Room to Move*, they are no less playful or worldly for it.[13] Indeed, this beauty is itself intimately linked to the ethical sociality that binds her to her dancers and audiences. Like metaphysics, beauty is, in the words of Iris Murdoch, 'unselfing' (in Scarry, 1999: 113). Elaine Scarry characterizes this 'unselfing' effect of the beautiful as a 'radical decentering,' a deep, somatic transformation that relativizes the ego in a manner reminiscent of Hae's admonition to 'think like artists. You have responsibilities to the Universe' (1999: 111, 130). For Scarry, an ethical component, indeed a praxical and daily one, is intrinsic to beauty: 'How one walks through the world, the endless small adjustments of balance, is affected by the shifting weight of beautiful things' (1999: 15). For Hae Kyung Lee and her dancers, the beauty of their final products together is the result of a larger and continuous process of unselfing in art and life, one in which 'the beautiful thing' is subsumed within ongoing ethical relations between self and others. And, if the 'endless small adjustments of balance' are organized and supported by this metaphysical sociality of ethics and beauty, so too are the seismic shocks.

'. . . like the words I cannot find, beyond words. . . ': dancing the ethical transformation of grief[14]

If dance and metaphysics are often partners, then death and metaphysics are intimates. In death we confront the ultimate alterity, the ultimate limits of referentiality, and, moreover, the ultimate challenge to an ethics of presence and response-ability. As Derrida reflected, in Levinasian terms, on Levinas's own passing: 'Death: not, first of all, annihilation, nonbeing or nothingness, but a certain experience for the survivor of the "without-response"'(2001: 203). de Certeau notes, after Foucault: 'All discourse finds its law in death, "the innocent good earth beneath the lawn of words"' (1986: 174). If, as both de Certeau and Derrida suggest, the seismic ruptures of death and loss are 'like the words I cannot find, beyond words,' they are also, and equally, beyond technique.

On Monday, 16 July 2001, Jill Yip, 28 years old, refugee from Vietnam, principal advisor for the College of Arts and Letters at California State

University at Los Angeles and a core member of Hae Kyung Lee and Dancers, was admitted to Garfield Medical Center in Monterey Park. Her intestine had blocked, then ruptured, flooding her body with a staph infection. She was in a coma from which she would not awake. For a week, Hae, Claudia and Miguel, and many of us who knew and loved Jill, moved in and out of the ICU, in and out of incomprehension, in and out of the fraught intimacy and utter corporeality forged at the intersection of shock and medical apparatus that seemed to both demand and foreclose present-minded focus. These were extraordinarily physical transactions. We rubbed her feet, her legs, her arms and fingers. We tried to talk her back but more, we tried, in our own ways, to dance her back or, even more accurately, at the level of musculature we tried to rub technique back into her body so she could dance back to us.

It was not to be. Late on the night of Monday, 23 July, Jill's family disconnected life support. Then, during the two days of memorial services, more extraordinary unwinding of corporeality from the technique that linked us as a community to our sense of Jill: the Chinese-speaking staff of the Universal Chung Wah Funeral Home, surveying the dozens of black, brown and white faces, enlisted translators to tell us what to do. They physically turned us into place, put the incense in our hands, routed us as required by the protocols of a Chinese Buddhist funeral, a different mode of performance than that we shared together.

Death and dance are both radically corporeal events that point us toward deeply social confrontations with the limits of representation, with, as Peters put it, 'the holiness and wretchedness of our finitude' (1999: 271). Yet, in the process of these social and personal confrontations lies an ethical potential both created anew and, in this community, rediscovered. Jane Bennett suggests that, in order to access this potential, it is necessary

> that one accede to what one cannot understand. Such untranquil acceptance contributes to a positive ethical result (i.e., the 'unworking of human arrogance'). This is the arrogance that purports to know the Other when, in fact, it is merely referring the Other back to some knowable version of the Same. To unwork such arrogance is to confess to the element of ineradicable difference that is the Other.... The potential for ethical respect lies within acceptance of finitude because 'the first experience of an alterity that cannot be reduced to the self occurs in the relation to death.'   (2001: 76)

Here, the alterity of 'the universe' and the 'unselfing' effects of beauty find their analogue in the 'ineradicable difference' of death. Perhaps it is no surprise, then, that the porous and nonprescriptive metaphysics so useful in accounting for antinomies in choreography, so central to constellating community, would become the 'good earth' under-girding communal grief. This metaphysics sustained company members by continually linking the pain and 'untranquil acceptance' of their loss to larger imperatives to care for each other and their work together. Paradoxically, the transcendent, by definition limitless, formulations that bind this group ('universe,' spirit, energy) became tactics to reckon with corporeal limits. These corporeal limits, in turn, offered, in Peters's terms, 'the closest thing to a guarantee across the chasm,' across the gulf separating a knowing beyond words from the daily disciplines of body and spirit.

One month after Jill's death I went to the dance studio at CSULA to watch Claudia and Miguel, Edgar Ovando and Jinna Rohde rehearse Hae's full length piece *Shadows of the Spirit*, which would be performed on 10 November 2001 at Los Angeles's Japan America Theatre. This was the piece Jill was rehearsing when she fell ill. It was already an elegy, dedicated to the memory of Duane Ebihara, Executive Producer of the JAT, a young man who also died suddenly and too soon the same year. Now this performance involved a double mourning.

The four dancers were on a trampoline, flinging themselves across, over and under each other. I was terrified; my protectiveness from our collective crisis had not worn off. Indeed, during the actual performance, audience members gasped and covered their eyes. Yet, both in rehearsal and in performance, these dancers did not miss. They were jumping so high, so fast, falling so close. They just know where they are, they told me later, adding that this sensation was hard to describe, like being careful without thinking. Like knowing and feeling somehow beyond words. Claudia and Miguel were smiling. They know each other's bodies so well they can feel each other, even in free fall, and their deep corporeal familiarity, their knowing and feeling beyond words, seemed to catch the other dancers too. This was a dangerous section in the middle of a long and exhausting dance but their complete presence to technique and to each other caught them, supported them, set them down gently.

This knowing, this familiarity – beyond consciousness and language, across difference, through crises – these are the daily small, good things which call so eloquently for the intimacy sustained by this group's ethics of service, and by the metaphysical performatives undergirding that ethic. In the generative sociality these metaphysical performatives organize and support, technique is mobilized in service of something more than expression, more than physical prowess, more, even, than beauty. What is mobilized here and circulated through 'tissues of bodies and tissues of language' (Oliver, 2001: 223) is a gentle, response-able way to be together as students, as artists and friends, as human beings who can imagine and incarnate ethical possibilities for living in, and creating, the world.

Conclusion: Dancing Communities – Ideas of Order, Queer Intimacies, Civic Infrastructure

In his poem 'The Idea of Order at Key West,' Wallace Stevens imagines the power of the solo performer to remake the world for herself and her audience. The afterlives of her performance arrange, deepen and enchant the material world and others' abilities to perceive it. It is as if the artist generates and inspires an irresistible order that is both incisive and evanescent: 'in ghostlier demarcations, keener sounds' (1982: 130).

Helen Vendler points out the 'endangering sentimentality' of this poem, noting that its uncritical celebration of the autonomous creator is itself unsettled by Stevens's own 'uneasiness' in his anxious piling up of superlatives and comparatives (1986: 68, 70). Yet, even with these cautions and qualifications, 'the maker's rage' this poem both demonstrates and describes compels the reader with the lyrical precision of its rhythms and imagery. Stevens's intentions aside, it offers what Diane Sippl has called the Brechtian 'pleasure of knowing the world can be remade' by creative work. Further, the poem suggests that such work can 'remake the pleasure of knowing the world' (*How Nice to See You Alive*, 1989: 84–5).

Writing from a very different perspective than Stevens, Randy Martin also argues for the performer as paradigmatic world-maker. He writes, in explicitly Brechtian terms:

Politics concerns the forces that devise the social world. The collision and mutual displacement of forces – their motional flows – is what makes for difference, and difference can be summed up, organized, and contextualized in myriad ways that produce a given society and structure its divisions along lines of class, gender, sexuality, race, and much more .... Politics goes nowhere without movement.

. . . The presumption of bodies already in motion, what dance takes as its normative condition, could bridge the various splits between mind and body, subject and object, and process and structure that have been so difficult for understandings of social life to negotiate.

(Martin, 1998: 3)

Here, too, performance has the potential to remake the world and our abilities to know and theorize it. Martin persuasively argues that dance's capacity to mobilize performers and audiences offers us an embodied model of politics in motion.

The daily work of art, not as a product but as routine labor, undergirds both Stevens's autonomous creator organizing the world and Martin's politicized *corps* changing the world. However monumental or moving art's final products, these are ultimately unintelligible outside of the physical and social routines that produce and surround them. In the case of dance, these routines are themselves enabled by technique.

Dance techniques are ideas of order, performative templates for generating artifice in/and community. They offer vocabularies for writing, reading, speaking and reproducing bodies. In doing this, they do much more: organize communities around common idioms, rewrite space and time in their own images, provide alibis, escape clauses, sometimes traps, sometimes provisional utopias. Beyond templates for generating artifice, they are also templates for arranging, deepening, and enchanting communities in and through the ghostlier demarcations and keener sounds of practice, rehearsal, performance. Techniques mobilize their object-bodies and their spectators by reproducing 'bunheads,' 'Apsaras' and vicarious romances. Sometimes they mobilize through ancillary discourses they inspire or require, like those of memory or metaphysics.

In 'Thirteen Ways of Looking at *Choreographing Writing*,' itself an allusion to another of Wallace Stevens's poems about the capacities of perception and artifice to remake the world, Peggy Phelan reflects on the ghostlier demarcations of bodies in performance. 'The moving body is always fading from our eyes,' she writes. 'Historical bodies and bodies moving on stage fascinate us because they fade' (Phelan, 1995: 200). Yet, for the 'moving bodies' themselves, the situation may be very different. Muscle has memory, as do minds and hearts. For the Sem family, historical bodies refused to fade in spite of atrocity, poverty and despair, even as they refused to be summoned despite an urgency linked to survival itself. For Hae Kyung Lee and her dancers, Jill Yip's body did not fade. It hovers in collective memory near steps in the repertory that must now be danced by another; it clings to choreography that needed to be reconfigured and, in so doing, remembered as she danced it. Likewise, despite requiems for concert dance that continually mourn dwindling audiences and diminished relevance, despite the paltry sums available to support dance as a public value and practice in this country and in Los Angeles (see Segal, 2004), dance itself refuses to fade. As of this writing, Robyn Gardenhire's dream of a ballet company that looks like Los Angeles continues to materialize. Oguri continues to organize dance caravans to explore the limits of the body in space. Young men and women, and old ones, continue to take class at Le Studio and train with Hae Kyung Lee at CSULA, incarnating what Wayne Booth calls the novice artist's 'paradox of hopeless hope': 'For any of us who did not start when young, the only imaginable rung on the ladder is – well, shall we say about 3 on a scale of 1–10? An approving smile from a teacher? What, then, can be the comfort?' (Booth, 1999: 113).

The comforts of dance technique are not simple ones. As the Sems demonstrate, there is sometimes no, or little, ultimate comfort to be had at all. Perhaps dance, and dancing bodies, refuse to fade from mind and heart, from private longing, amateur practice and public performance because technique, in all its various incarnations, offers not (or not only) comfort, but a kind of queer intimacy characterized by what Lauren Berlant and Michael Warner call 'a space of entrances, exits, unsystematized lines of acquaintance, projected horizons, typifying examples, alternative routes, blockages, incommensurate geographies' (1998: 558). Bodies and identities are remade in dance technique's queer intimacies; within them, as Martin observes, the 'diverse capacities and practical differences that compose a self are assembled' (1998: 179). But more than this, the queer intimacies of technique build and support complex socialities by bridging the binaries of present versus past, the utterly impersonal versus the radically local, and individual self-expression versus group solidarity as they rewrite bodies, space, time and relationships. To be sure, not all of the socialities thus formed are liberatory or even generative. Some offer their members models for remaking the world based on an embodied ethics of response-ability, whether explicitly invoked or not. Some confront personal and social despair; some succumb to them. Some offer fleeting and vicarious intimacies, others lengthy and sustaining ones.

Dance technique is consumed, resisted, deployed and redeployed like any other aspect of culture. Whatever its elite allegiances or its resistant ambitions, technique is also routine, ordinary, a series of tactics for living, not simply a strategy for moving. To adopt this view is to move beyond a cramped sense of dance as artifact and toward a more nuanced understanding of how it actually serves people who make and consume it. This view offers a way of examining the rhetorical and social by-products of performance, not as mere incidental sources of pleasure or skills – humanist gilding on the aesthetic lily – but as self- and world-making ventures central to a politicized, engaged approach to cultural forms. It takes the affective as well as the physical labors of creation seriously. This view of technique as a template organizing and mobilizing complex intimacies and communities argues for even closer examinations of the ways the daily operations of performance expose, manage, finesse, evade, and often transform the tensions, constraints and opportunities that must be continually negotiated by embodied subjects within the global city.

What does it mean to take the self- and world-making powers of performance seriously in urban analysis and urban planning? To start with, it means engaging the ways community and citizenship are performed in and through everyday embodied practices of cultural production and consumption. It means rethinking space and place, the literal grounds of urban experience, as performatively produced through embodied engagements, and not taking them for inert 'givens.' It means public recognition of 'ethico-aesthetic invention' as a source of renewal for urban politics (Amin and Thrift, 2002: 158). Shannon Jackson examines precisely these issues in her analysis of Hull House. Here, the daili-ness of performance reshaped both the participants' aesthetics and their civic attachments. Such reshaping was an explicit goal of settlement 'reformance' (2000: 212–13).

Globalization is frequently analyzed in terms of various kinds of flows: the flow of capital, of telecommunications, of bodies. Yet the global city is also a site of concentration, not simply a nodal point in a larger network of production and dispersal. As Saskia Sassen observes:

> Alongside the well-documented spatial dispersal of economic activ-ities, we are seeing the growth of new forms of territorial centraliza-tion in top-level management and control operations.... Once these processes are brought into the analysis, funny things happen; secret-aries become part of it, and so do the cleaners of the buildings where professionals work.   (2000: 1)

This concentration of capital and bodies offers possibilities for other 'funny things' to happen in ballet studios and in performance venues: solidarity with 'the guys' or 'my girls,' vicarious romances with monsters, intimacy in/as corporeal 'reform' with constituencies including the cleaners, secretaries and managers simultaneously. Here it becomes possible to rethink the very nature and force of social capital, and to reconceive sites of connections between bodies in the global city across multiple dimensions of difference.

All global cities offer similar opportunities to dancing communities, at least in the abstract: rich juxtapositions and hybrids – the very preconditions for queer intimacies – borne of concentration, diversity and cultural infrastructure. Thus, they are natural incubators of cross-cultural and transnational forms. But Los Angeles frames this potential in unique and specific ways. Its own self-rejuvenating amnesia may be as much popular myth as civic ethos but, in either form, it can be generative. No single canonical form, artist, company or tradition can stake an unassailable claim on civic culture using appeals to real or imagined consensual history. Here invention trumps history. Indeed, LA's civic ethos and popular mythos converge on precisely this point, with and without nostalgia or irony.

Further, the demographic and aesthetic realities of dancing communities in Los Angeles expose fictions of a coherent 'American' culture in the city so intimately associated with disseminating it. Here dancers, choreographers and teachers poach on, engage, resist or simply ignore the entertainment industry and its aesthetic vocabulary. In so doing, they reveal the partiality of, and erasures from, the industry's products, even as they reflect, on a small scale, the fact that, for decades, 'Hollywood' and the Americana it creates, enables and distributes 'has been put on a global auction block' (Davis, 1992: 143). 'American' culture in Los Angeles is a commodity with centripetal force, pulling toward an imagined center: the largest possible market. Dancing communities here do not speak with one voice. Some want to reproduce classical techniques and others reject them, to take only one example. Yet each offers its members two common opportunities: productive interpersonal friction and connection, organized by technique, within and across difference; and domestication of the urban landscape in terms other than those of market forces.

The dancing communities described in this book attest to the fact that productive, even utopian models for reimagining urban life are all around us. The Sems may have been unsuccessful at using dance to answer back to terror but Sophiline Shapiro continues to bind the global to the local, and the past to the present, as she teaches Khmer classical dance to a new generation in Long Beach. Students at Le Studio continue to redeploy ballet to their own ends as they grow up together in solidarity and in difference.

Imagine the innovative, generative social arrangements that could emerge if there was the political will to seed and support sites of art-making as laboratories for complex, productive forms of citizenship on a large scale. Imagine the ethico-aesthetic equivalent of corporate industrial parks – corporeal art parks – sites of communal practice and public performance that produce diverse affiliations and vicarious identifications, evolving civic landscapes, enabling infrastructure for memory, enchantment, generosity, all forged through the daily routines of moving together. Imagine arts education, even in-depth apprentice-ships, required of all students because they serve the public good as well as individual talent.

In *Friction: An Ethnography of Global Connection*, Anna Lowenhaupt Tsing observes: 'Global connections are made in fragments – although some fragments are more powerful than others' (2005: 271). This is true within the global city itself, as well as in larger relationships between such cities and globalization's margins. Tsing adds that a focus on such fragments 'need not reduce analysis to simply noticing idiosyncrasy and happenstance' (2005: 271).

The fragments of generative, even intimate, interpersonal and inter-cultural connections I've discussed here are both idiosyncratic and part of a much larger repertoire, one linking multiple, unique places and histories, one where performers, their teachers and their audiences find and make communities in solidarity and in difference from national venues to cement-floored garages. Are these templates for sociality, these poetics for self- and community fashioning, sustainable, replicable? In Venice and Pasadena and Long Beach, on caravans through the desert and in a studio on the border of East Los Angeles, and in countless other venues grander and more modest, dancers, poised between one step and the next, reach out to us as partners, eager to show us the next move.

Notes

Introduction: dancing the city

1. In the most reductive terms, a progressive politics of propinquity postulates 'that civic culture will emerge from mingling in public spaces' (Amin and Thrift, 2002: 137). For a more complex view, see Copjec and Sorkin, 1999.
2. Unlike many contemporary scholars, performance historians often explicitly link the economic-demographic aspects and the everyday dynamics of urban art-making in their analyses. Especially noteworthy in this regard is Shannon Jackson's *Lines of Activity*, a study of the role of performance in/as 'civic housekeeping' in Jane Adams's Hull House.
3. This definition of 'neighborhood' is loosely based on Mayol, 1998: 10–11.
4. See Ness (1996) for a discussion of some exemplary failures of these approaches.
5. See also Taylor, 2003: especially pp. 19–21.
6. Founder Joseph Rickard named his company the 'First Negro Classic Ballet,' though it was not, in fact, the first. In 1937, the American Negro Ballet debuted in Harlem. See Blodgett and Hodgson, 1996; and Horwitz, 2002: 321.
7. Edward Soja uses the formulation 'thirdspace' to characterize 'a limitless composition of lifeworlds that are radically open and openly radicalizable' (1996: 70). 'Thirdspace' proceeds from the ontological assumption of 'trialectics,' which identifies and links spatiality, history and sociality as central to knowledge production. The choreographers I refer to rework the here and now of contemporary dance vocabularies in light of traditional or classical ones from there and then to generate a new technique and performance community. Hae Kyung Lee, discussed in Chapter 4, is one such artist.
8. Amin and Thrift (2002) see these two modes of connection as opposites. A performance-centered social aesthetics suggests that they are, more accurately, productive partners.

Chapter 1   Intimacies in motion

1. While dancing communities create and sustain many of the conditions necessary for the queer intimacies Berlant and Warner describe, it is impossible to deny the heteronormative couplings, real and vicarious, central to this chapter and, indeed, to much of the classical Western repertory. The use of Irigaray later in the chapter is also problematic with respect to the arguments of Berlant and Warner; further explication of this is given in note 4 below. In spite of these points of disagreement, I argue that the intimacy plots sustaining dancing communities are expansive and replete with queer possibilities. For additional examples, see Desmond, 2001.
2. No limiting dyad is intended by the examples presented here. The vicarious romances forged by spectatorship are just as likely to emerge in the routine

transactions of training. And virtuoso performances certainly bind artists to audiences by the routine formulations used to characterize extraordinary performance in everyday conversations.

3. Taylor writes: 'The scenario includes features well theorized in literary analysis, such as narrative and plot, but demands... attention to milieux and corporeal behaviors...' (2003: 28).

4. Peters's argument involves the idea of communication generally and epistemologically; interpersonal exchange is one component of this.

5. As I noted in 'I Dance to You: Reflections on Irigaray's *I Love To You* in Pilates and Virtuosity,' and below as well, Irigaray oscillates between essential and metaphysical invocations of the body in her text and, in so doing, is heterosexist in her presumptions. See 1996: 230–1. For critiques of these presumptions, see also Eric Doxtader's 'Loving History's Fate, Perverting the Beautiful Soul: Scenes of Felicity's Potential,' and Craig Gingrich Philbrook's 'Love's Excluded Subjects: Staging Irigaray's Heteronormative Essentialism,' *Cultural Studies* 15.2 (April 2001): 206–21 and 222–8.

6. Certification in a newly popularized and rapidly expanding technique like Pilates is not a centrally regulated process. There are multiple 'certifying' institutions. This had been complicated further by the fact that, until October 2000, Pilates technique was itself a legally contested site. Joseph Pilates never copyrighted or trademarked his procedures. Many of the dancers with whom he worked went on to serve as instructors following the equivalent of informal apprenticeships. After his death, the owner of a Manhattan studio moved to register 'Pilates' as a trademark and to strictly regulate its use, including pursuing legal action against Pilates' former pupils who had been using the name for decades. Ultimately, the trademark was not upheld, and practitioners and certifying entities were free to advertise "without skirting the name" (see Morse, 2001: S8). The staff of Le Studio Fitness is certified by On Center Conditioning after 100 hours of academic study (including presentation of case histories), 100 hours each of work on the apparatus and observation, and 200 hours of teacher training for a total of 500 hours.

7. See also, for examples, Browning, 1995; Morris, 1991: 3; and Desmond, 1997: 2.

8. Doxtader is arguing specifically for a field of play in an ethics of carnal desire. I am extending his insights to include such a field of play for desire in/as spectatorship. Doxtader clearly states, and I agree, that such fields, while liberatory, are not utopian in any simple way (2001: 217).

9. See Doxtader, 2001: 220, n. 15.

10. Sippl's analysis is a close reading of five films. Her general argument, however, is remarkably and compellingly relevant to dance, and to the video performances I discuss below.

11. Palmer argues for the potential of the virtuoso to be deployed rhetorically in the construction of Victor Turner's 'communitas.' Certainly dancing communities offer possibilities for such collective, utopian moments in performance. But my concern here is with the specific interpersonal, affective work organized by technique and inspired by the virtuoso.

12. Oguri no longer uses his first name, Naoyuki, as he did in his early Los Angeles performances.

13. My position here is in accord with Amelia Jones's argument about the dynamics of spectatorship in body art. She writes: '... I stress again here that the presentation of the body/self in body art marks not the immediacy, unity, and presence of this body/self but its radical interdependence with the other. This is so, crucially, *whether or not body art's phenomenological disturbance is experienced "in the flesh" or through visual or textual documentation'* (1998: 107; emphasis in original).

14. Schneider is particularly interested in the 'art–pornography' divide, and in 'who gets to make what explicit where and from whom' (1997: 20). She focuses on feminist performance artists who deal explicitly with gender, race and sexuality. While Oguri's work challenges conventional categories of difference, it does so in more oblique ways than many of the pieces Schneider describes.

15. Lyotard's definition of the sublime is actually a reading and reapplication of Edmund Burke.

Chapter 2 Corporeal chronotopes: making place and keeping time in ballet

1. Over the course of my research, Le Studio's resident company has changed. In June, 1996, Le Studio's 16-year relationship with the Pasadena Dance Theatre (PDT), its then resident company, was terminated by mutual agreement of the Fullers and the PDT Board. See correspondence between Charles Fuller (23 April 1996), James Aguiar (16 May 1996), and Charles and Philip Fuller (26 May 1996) in the Le Studio newsletter ('News Flash,' n.d.). Dance Corps was constituted and founded by the Fullers in 1996. There was considerable continuity of membership from PDT to DanceCorps; the major factor here, in my determination, was loyalty to the Fullers and to Le Studio as a training facility.

2. Christensen, or 'Mr C' as the Fullers affectionately referred to him, founded Ballet West in 1968. He was the first to stage complete productions of *Coppélia* (1939), *Swan Lake* (1940), and *The Nutcracker* (1944) in the United States. He passed away in October 2001 at the age of 99. See Anderson, 2001.

3. With the shift in affiliated companies, the *Nutcracker* performance also changed. With the earlier company, a full-length version of the ballet was performed. Some of the informants quoted here are referring to this production and some to the hour long *The Story of the Nutcracker*; virtually all of the major variations are included in the latter. In addition, Le Studio is the common denominator in both performances; both casts were composed of Le Studio students.

4. At Le Studio, students move through four general divisions: Threshold (including pre-ballet, for students between 3 and 7 years old); Basic (with levels I, II and III), 'Tech' ('Technique,' with levels I and II) and Advanced. The timing of advancement generally parallels the academic school year. Assessment of students' progress and physical development happens in the spring, with the new level taking effect, generally, in the fall.

5. Kirkland characterizes her departure from the New York City Ballet as a 'defection,' one of a number of Cold War metaphors in her memoir (1986: 113).

6. Though this was not mentioned in the body of the article, DelaCerna, in a letter to the magazine over a month later, stated that all Misty's earnings were deposited in an account to be used only by and for her daughter. See DelaCerna, 2000.

7. As of this writing, Misty Copeland is a member of the American Ballet Theatre.

8. Ballet dancer Frank Martinez is not to be confused with Francisco Martinez, another well-known Los Angeles dancer and founder of the Francisco Martinez Dance Theater.

9. For a New York version of a similar project, see Dunning, 2001. Robin Gardenhire and The City Ballet of Los Angeles continue to perform and offer classes at the Red Shield Center. In 2005, the company was also invited to start a new program at the Salvation Army in South Central Los Angeles.

10. Years after this interview, Erica's positive attitude and her commitment to ballet were shared with a national audience. In an article on the 2003 New York International Ballet Competition, Shattuck details the physical and emotional pressures facing the competitors. The article ends with a paragraph quote from Erica, who attests to the kind of generative power of working with other dancers (Shattuck, 2003: AR27); her attitude here is utterly reminiscent of her sense of the solidarity in competition forged at Le Studio.

11. While these parents were quite vocal about their concerns, they were more reticent about consenting to have their real names used. In part, I suspect, they did not want to appear unsupportive of their daughters' ambitions, even if they wished these ambitions were more 'conventional.' 'Mrs A,' who reassured me that 'of course' she was very proud of her daughter, was a particularly interesting case. She seemed anxiously poised between the role of ardent stage mother, critiquing her daughter's and other girls' performances as we watched company class, and a more ambivalent position of fretting that ballet would 'put [her daughter.] out of the running' for both an 'A-1 college' and social position.

12. In *Distinctions*, Bourdieu argues that the 'highly ambivalent relations' between artists and patrons of the arts may lead artists to find 'structural homology' between them and the dominated classes' (1984: 316). Moreover, he argues that the patrons of the arts respond to artists with a paternalism 'in the name of a not-so-unrealistic image of what the producers of cultural goods really are, that is, deviant children of the bourgeoisie or "poor relations" forced into alternative trajectories; the patrons may even find a pretext for their exploitation of the artists in their conspicuous concern to protect them from the consequences of their "idealism" and their lack of "practical sense"' (Bourdieu, 1984: 316). While Bourdieu's reading may be accurate in its broadest contours, it fails to take account of the even greater disparity between artists and patrons in Western European countries where dancers, for example, have health care and pensions even if they are not with 'star' companies, and those in the United States where the status of artists other than 'stars' is much lower materially and socially. I recall a conversation in the Le Studio dressing room where I was asking Gilma Bustillo, who had danced with several companies in Western Europe, about the economics of ballet employment there. She was

talking about her medical plan and vacations, and her pension; a woman who worked as a freelance graphic designer, suiting up to take the adult class, turned to her friend, also a freelancer, and said, 'To think my folks steered me away from dance and into college because they said it wasn't secure.'

13. One aspect of the body that is officially and explicitly *not* controlled at Le Studio is weight. Girls across four 'generations' of company members attest to this. 'It's not an issue,' Erica noted firmly when asked. Indeed, a number of Tech II girls and some company members across the years have been heavy, not only by ballet standards but by everyday ones. 'It didn't get mentioned,' observed 'Jennie,' a recent 'graduate' from Le Studio who has moved on to another regional company. 'The guys [Chip and Phip.] were always on us to eat healthy.'

   This bears mentioning as accounts of anorexia in ballet circulate through both insider discourse at many studios and popular media, leaving the impression that the bony dancer favored by Balanchine is automatically synonymous with ballet across all amateur and professional contexts. To be fair, in some studios girls are put on notice that weight is an issue, either through explicit pronouncements or through indirect means like not getting cast or being cast in character parts. In my own early ballet training, pressure to stay thin was pervasive and relentless. At Le Studio, however, this is not the case. This is not to say that the girls themselves are unaware of weight; they are. Still, they appreciate that, in the words of 'Lisa,' 'It's [watching one's weight/dieting.] something we do if we want to and some do and some don't. It's not like they say "Don't eat those fries!" We pretty much eat what we want.' She goes on to describe the one time she did hear weight mentioned, when 'Michael,' a high school senior and company member who began training at Le Studio when he was 7 years old, was preparing for a ballet competition; he was a finalist in this contest and would perform a solo at the Dorothy Chandler Pavilion in downtown Los Angeles. 'Lisa' recalled, 'Phip was like, "Caesar salads, buddy, Caesar salads," cause ['Michael'.] was, you know, not fat but, like, big then and we [the company girls.] were like, "Yah. Right. That's a joke. Do you know how many calories are *in* a Caesar salad? Like about ten million!"'

14. Pointe shoes are not actually 'blocks,' though, as Jennifer Fisher notes in 'Pointe Shoe Confidential,' some assume the shoes have wooden blocks in the toes. This assumption is based on a folkloric corruption of an aspect of the manufacturing process: 'the toe "boxes" are made of layers of fabric and paper, held together with glue and are shaped by hand before they harden. They are molded on fiberglass lasts (forms in the shape of a foot), but the lasts used to be made of wood, making the shoes "wooden blocked"' (Fisher, 1999: 64).

15. Even 'The Trocks' attest, indirectly, to the intense discomfort of pointe work. As Fernando Medina Gallego ('Svetlana Lofatkina') observes: 'When you are classically trained and you see girls doing it, you think, "Well, it doesn't seem so difficult." Then you get the pointe shoes on, and you know otherwise. When I'm dancing in a tutu, I imagine now that I have to look like a ballerina. But what is telling me I'm a ballerina is the pain in my feet' (in Shattuck, 2002: AR7)

16. As noted above, an increase in boys enrolled in earlier divisions suggests that is may be changing.

17. This motif of ballet as a way of 'keeping kids off the streets' seems to recur as a trope across multiple accounts. Erica's narrative is one example. Another is an early account of current international ballet star Carlos Acosta, wherein his father, '... Pedro Acosta, a truck driver, decided to keep his 9-year-old son out of trouble by virtually forcing him into a state-subsidized ballet school in Havana' (Kisselgoff, 2002).

18. The men may have left the stage but they didn't leave the theatre. They continued – and continue – to dominate the backstage infrastructure of ballet as choreographers, ballet masters, artistic directors, and producers. See Hanna, 1988: 126.

19. Burt argues that Vaslav Nijinsky (1889–1950) was a special case who could claim the category of 'genius' during this period. But, in many ways, he was the exception that proved the rule.

20. The analogies may, occasionally, also run the other way; in 'Locker Room Talk,' George Garrett describes a young man he played against in a neighborhood game: 'Watching that guy run with a football was like watching Baryshnikov at his best' (1992: 117).

Chapter 3   'Saving' Khmer classical dance in Long Beach

1. Photographs were not an option for this chapter. After years of repression, the Sem family, and Ben in particular, still felt they were living under threat. They wanted no images or recordings made of their dance exercises or interactions with me. Ben was very nervous about my even taking notes. This reticence was understandable in light of the Khmer Rouge practices discussed later in this chapter.

2. Long Beach, California, is home to a population of up to 50,000 Cambodians, the overwhelming majority of whom arrived after 1980. The Anaheim Street neighborhood has been unofficially designated 'Little Phnom Penh.' See Wride; and Pearlstone, 1990: 83–4. It has deep and activist ties to Cambodia that include the cultural and the political. See, for example, Decherd, 2001; Pape; and Mydans, 2000.

3. For example, the name of the former Director of the Department of Dance at the University of Fine Arts in Phnom Penh is sometimes transcribed as 'Pich Tum Kravel.' In other places it is spelled 'Pech Tum Kravel,' suggesting, to English speakers, subtle but significant differences in pronunciation. See Dy in Ebihara, 1994: 30.

4. There were at least two other strong possibilities. One was the Sihanouk-affiliated Site B, and the other was the much smaller Sakeo II camp. Camp identification is a significant matter for, as noted below, Cambodian factions were assigned to specific camps. Site 2 was administered by the Cambodian People's National Liberation Front, a non-communist, non-Sihanoukist group which was itself extremely fragmented. See Hall (1992) for a map of these camps.

5. May's emphasis on what she viewed as performances of the reading and writing of charges by those she characterized as low-level Khmer Rouge soldiers seems fairly unique. The majority of published survivors' accounts recall oral, seemingly improvised performances of such charges, which heightened

the sense of their arbitrariness. Orality, too, is more in line with the nonliterate peasantry that, in most accounts, formed the core of Khmer Rouge support. One exception to the dearth of references to writing can be found in Vann Nath's account, in which he describes the chief warden of Tuol Sleng reading his 'biography,' as well as written letters ordering arrests (1998: 49, 27). As Ben Kiernan points out in *The Pol Pot Regime*, the Khmer Rouge was not homogenous in either its membership or its methods. One particular context in which writing does emerge as intrinsic to Khmer Rouge terror is interrogation. See, for example, Nath, 1998: 32.

6. To be fair, Sophiline Cheam Shapiro is of a younger generation than Ben and May Sem, though this in no way mitigates the horrors of her own experience with the Khmer Rouge; she was 8 when she and her family were forced from their homes. She also had much more extensive formal training as a member of the first class to graduate from the reconstituted School of Fine Arts and of the National Dance Company of Cambodia. See her account in Pran, 1997.

7. Relative mastery of English figured in another gendered divide of the Sem household. Both Sandy and Jennie were very verbal and confident; they were fairly fluent English speakers, though they did have some difficulties reading and spelling the language. Rith's difficulties with reading were more significant. Further, over the time of our acquaintance, his speech seemed to actually regress. He moved from an exuberant, almost boisterous, delight with spoken English to a reluctance to speak at all. Toward the end of our acquaintance, he communicated in sharp, staccato bursts of Khmer with occasional interjections of English.

Because May spent so much time with her girls, including putting up with their sometimes relentless corrections, her English skills were much more developed than her husband's. This was particularly noticeable in her mastery of relative subtleties like articles, tense and vocabulary. Though she was not a sophisticated speaker by formal, academic standards, she was certainly clear by the norms of everyday communication. Ben's isolation from family and community was reflected in his relatively poorer facility with English which, in a vicious discursive circle, isolated him still further from his daughters, who did not dare to correct him. It seemed to me, too, that Ben resented this gendered fluency. Early in my visits when he would search for an English word ('Say –?'), he would pre-emptorily brush aside May's and the girls' attempts to translate and pointedly turn to Rith. Of course this became a less and less viable option. Once, as he prepared to leave the apartment, May asked him to stay; I made out the equivalent of 'learn the English.' He snorted, replied sharply, and left. Jennie, who had witnessed the scene, told me reassuringly, 'It's not you. He said he hates English in his ears.'

8. The Khmer community in Long Beach recognized that spousal and child abuse resulting from posttraumatic stress was a serious problem. By 1997, Cambodians made up over 30 percent of the child abuse cases referred to the Asian Pacific Project of the Los Angeles Department of Children and Family Services; traumatized parents saw confinement and whipping as viable options for controlling children and keeping them from gangs (J. Stewart, 1997: B3).

9. See Phim and Thompson, 1999: 38. See also their discussions of the institutionalizations of the *Lakhon Khol* by King Norodom Sihanouk (1999: 54–6),

and the establishment of a 'folk tradition' section, with accompanying royal support, at the University of Fine Arts in Phnom Penh (73; also Charlé, 2001).

10. Norodom Sihanouk's royal career is complex. He was crowned King in 1941, then abdicated in favor of his father in 1955 to serve as both Prince and Head of State. He was crowned King again in 1993.

11. Cambodia is a postcolonial site, having been granted independence from France in 1954.

Chapter 4   Dancing other-wise: ethics, difference and transcendence in Hae Kyung Lee and dancers

1. I draw this chapter title from Leonard C. Hawes's important essay 'Becoming-Other-Wise: Conversational Performance and the Politics of Experience' (1998) that is so central to my argument in Chapter 1.

2. Banes (1998) argues that spiritual rhetoric and imagery in modern dance, and in Wigman's case in particular, served a variety of social, political and economic functions. Her analysis is confined to the dances themselves, and to choreographers' public pronouncements. She does not discuss how such vocabularies organize dance communities on a daily basis apart from the exigencies of public performances and promotional materials. Barbara Browning does ask, 'What constitutes choreography when the design of motion is ascribed to divine sources?' (1995: xxiv) in her ethnographic study *Samba: Resistance in Motion*.

3. I am not trying to conflate de Certeau's 'mysticism' and Levinas's 'transcendence,' or their specific spiritual traditions and commitments here. Both, from their unique historical and philosophical projects, offer insights and vocabularies which are useful for (1) charting various conceptions of the relationship between the corporeal and the ineffable, and (2) formulating a description of ethics as it is constructed in Hae Kyung Lee and Dancers.

4. These words of de Certeau in the opening paragraphs of *The Mystic Fable* are polyvalent. Though I am characterizing his assertions as descriptive of the ineffable's exile/alienation from referentiality, and particularly from theoretical discourse, they may be read differently in light of his projects and his personal theological commitments. Ahearne suggests that they may refer to 'the irreducible existential separation which divides him as a historian from the object of his study' (1995: 96). They may also suggest the impossibility of any unproblematic relation with the Christian tradition so central to de Certeau's life; he was ordained a Jesuit in 1956 and never simply renounced his faith or the affiliation. That so many readings of these opening paragraphs can be accommodated attests to the productive elasticity that can arise in and around rhetorics of the ineffable, as I will discuss below.

5. This discussion of options for managing intrapersonal, spiritual experience is heavily indebted to Sells, 1994: particularly 2–3.

6. Unlike many at CSULA, including many dance majors, Claudia did have dance training before entering the university. She and her sisters performed with folklorico groups, though, as she pointed out, her participation was fairly limited and she never felt connected to the form.

7. See, for example, Isadora Duncan's condemnation of jazz and ragtime dance as 'the sensual convulsions of the Negro' (in Daly, 1997: 217). It may be argued that these utterances from the early twentieth century are not an accurate reflection of the technique now adopted by the Alvin Ailey Company among others but, as Franko observes, Duncan's recoding of movement, and her self-created myth of the dancer-subject, 'can be read as a foundational narrative of modern dance, as its myth of origin' (1995: 1).

8. For an account of another Los Angeles artist's ambivalence to multiculturalism, see Cheng, 2000.

9. While Hae's terms recall those of Bakhtin's in 'Art and Answerability' (see Chapter 3), the metaphysical performatives she relies on to characterize her ethics would be anathema to him. He explicitly discounts formulations like 'inspiration' or 'possession' (1990: 2) and centers answerability precisely on 'guilt or liability to blame' (1990: 1).

10. Dimock's larger argument deals with the shortcomings of a philosophical concept of justice as a syntactic grid of commensurabilities. In contrast, she offers a view of literature that transposes the 'clean abstractions of equivalence into the messiness of representation' (1996: 10). Her critique of commensurability resonates with Oliver's critique of recognition. See especially Oliver's account of incommensurability in the 'accuracies' of accounts by historians and a Holocaust survivor (2001: 1–2). Further, Dimock's sense of what 'gets lost in translation' in fantasies of commensurability (2001: 6) resonates with the view of the ineffable as a productive ambiguity that resists denotation.

11. Julio's path is typical of many CSULA student dancers. At the time of this performance, he was pursuing a biology degree with hopes of going to medical school. Jose Reynoso came to Hae's classes, and joined the company while completing his Masters degree in psychology at CSULA.

12. To be sure, there are also some clear differences between Coleridge's text and Hae's dance. The most substantive of these is that Coleridge's Mariner must atone for a personal offense, even as actual punishment is visited upon others. Likewise, his redemption is an individual matter. In characterizing her work as both inspired by and directed to 'the universe,' Hae relativizes the importance of any individual, including herself, and casts her choreography as a kind of purifying gift that rebounds between artists, audiences and the ineffable.

13. Relations between dance, particularly modern and postmodern dance, and beauty are complex and contested. Consider, as one symptom of this complexity, a symposium, 'What Dance Has to Say About Beauty,' moderated by Ann Daly, in which a number of curators and choreographers resisted talking about beauty at all, or dismissed it out of hand. For example, choreographer Michael Carson stated: 'I had Mary Wigman-based training. I'd always been taught that traditional beauty was something that you didn't really strive for.... If it was just a little bit too beautiful, it was somehow frivolous' (2000: AR26). Interestingly, Hae Kyung Lee was invited to serve as artist-choreographer in residence at the Mary Wigman studio (Palucca Schule) in Dresden in the Summer of 2002.

14. This quotation is taken from 'All-Out Friendship,' Jacques Derrida's essay on the death of Jean-François Lyotard. See *The Work of Mourning*, 2001: 215.

Works Cited

Abu-Lughod, Janet L. (1999). *New York, Chicago, Los Angeles*. Minneapolis: University of Minnesota Press.

Adato, Alison. (1999). 'Solo in the City.' *Los Angeles Times Magazine*, 5 December: 16+.

Ahearne, Jeremy. (1995). *Michel de Certeau: Interpretation and its Other*. Stanford, CA: Stanford University Press.

Amin, Ash and Nigel Thrift. (2002). *Cities: Reimagining the Urban*. Cambridge: Polity.

Anawalt, Sasha. (1996). *The Joffrey Ballet*. Chicago: University of Chicago Press.

Anderson, Benedict. (1991). *Imagined Communities*. London: Verso.

Anderson, Jack. (2001). 'Willam Christensen, 99, Dies.' *New York Times*. 17, October: A22.

——. 2003 'Young Performers with Something to Celebrate.' *New York Times*, 1 July: B5.

Aristotle. (1985). *Nichomachean Ethics*. Trans. T. Irwin. Indianapolis: Hackett.

Attridge, Derek. (1999). 'Innovation, Literature, Ethics: Relating to the Other.' *PMLA*, 114.1: 20–31.

Auslander, Philip. (1999). *Liveness: Performance in a Mediatized Culture*. New York: Routledge.

Babitz, Eve. (1994). 'Bodies and Souls.' In *Sex, Death and God in L.A.* Ed. David Reid. Berkeley: University of California Press, 108–50.

Bachelard, Gaston. (1964). *The Poetics of Space*. Trans. Maria Jolas. Boston: Beacon.

Bahti, Timothy. (1996). *Ends of the Lyric: Direction and Consequence in Western Poetry*. Baltimore: Johns Hopkins University Press.

Bakhtin, Mikhail. (1981). *The Dialogic Imagination*. Ed. M. Holquist. Trans. C. Emerson and M. Holquist. Austin: University of Texas Press.

——. (1990). *Art and Answerability: Early Philosophical Essays*. Ed. M. Holquist and V. Liapunov. Trans. V. Liapunov. Austin: University of Texas Press.

——. (1993). *Toward a Philosophy of the Act*. Ed. V. Liapunov and M. Holquist. Trans. V. Liapunov. Austin: University of Texas Press.

Banes, Sally. (1998). *Dancing Women: Female Bodies On Stage*. New York: Routledge.

Banes, Sally and Noël Carroll. (1997). 'Marriage and the Inhuman: *La Sylphide's* Narratives of Domesticity and Community.' In *Rethinking the Sylph*. Ed. Lynn Garafalo. Hanover, NH: Wesleyan University Press, 91–105.

Barthes, Roland. (1978). *A Lover's Discourse: Fragments*. Trans. Richard Howard. New York: Hill & Wang.

——. (1982). *Empire of Signs*. Trans. R. Howard. New York: Hill & Wang.

——. (1985a). 'The Grain of the Voice.' In *The Responsibility of Forms*. Trans. R. Howard. New York: Hill & Wang, 267–77.

——. (1985b). 'The Romantic Song.' In *The Responsibility of Forms*. Trans. Richard Howard. New York: Hill & Wang, 286–92.

Bataille, Georges. (1984). *The Absence of Myth*. Trans. Michael Richardson. London: Verso.

——. (1986). *Erotism.* Trans. Mary Dalwood. San Francisco: City Lights.

Battaglia, Debborah. (1995). 'On Practical Nostalgia: Self-Prospecting among Urban Trobrianders.' In *Rhetorics of Self-Making.* Ed. D. Battaglia. Berkeley: University of California Press, 77–96.

Becker, Gay and Yewoubdar Bevene. (1999). 'Narratives of age and uprootedness among Older Cambodian Refugees.' *Journal of Aging Studies,* 13.3: 295–314.

Bennett, Jane. (2001). *The Enchantment of Modern Life.* Princeton: Princeton University Press.

Bennett, Susan. (2001). 'Comment.' *Theatre Journal,* 53.2 (May): np.

Berger, John. (1992). *Ways of Seeing.* London: Penguin.

Berlant, Lauren and Michael Warner. (1998). 'Sex in Public.' *Critical Inquiry,* 24.2: 547–66.

Bhabha, Homi. (1996). 'Aura and Agora: On Negotiating Rapture and Speaking Between.' In *Negotiating Rapture: The Power of Art to Transform Lives.* Chicago: Museum of Contemporary Art, 8–17.

Billy Elliott. (2000). Dir. Stephen Daldry. Working Title Films and BBC Films.

Bishop, Elizabeth. (1984). 'Crusoe in England.' *The Complete Poems: 1927–1979.* New York: FSG, 162–6.

Blau, Herbert. (1982). *Take Up the Bodies: Theatre at the Vanishing Point.* Urbana: University of Illinois Press.

Blodgett, Peter J. and Sara S. Hodgson. (1996). 'Worlds of Leisure, Worlds of Grace: Recreation, Entertainment and the Arts in the California Experience.' *California History,* 75.1 (Spring): 68–83.

Booth, Wayne. (1999). *For the Love of It: Amateuring and its Rivals.* Chicago: University of Chicago Press.

Bourdieu, Pierre. (1984). *Distinction.* Trans. R. Nice. Cambridge, MA: Harvard University Press.

Briginshaw, Valerie A. (1997). '"Keep Your Great City Paris!" – The Lament of the Empress and other Women.' In *Dance in the City.* Ed. Helen Thomas. New York: St. Martin's, 35–49.

Brodsky, Joseph. (1994). 'Homage to Marcus Aurelius.' *Artes*: 38–55.

Brody, Liz. (1998). 'The Power of Pilates.' *Los Angeles Times.* 16 November: S1+.

Brookings Institution. 'Los Angeles in Focus: A Profile from Census (2000).' (2003). Washington, DC: Brookings Institution.

Brown, Linda Mikel and Carol Gilligan. (1992). *Meeting at the Crossroads.* New York: Ballantine.

Browning, Barbara. (1995). *Samba: Resistance in Motion.* Bloomington: Indiana University Press.

Brumberg, Joan Jacobs. (2000). 'When Girls Talk: What It Reveals about Them and Us.' *Chronicle of Higher Education.* Section 2. 24 November: B7–10.

Burt, Ramsay. (1995). *The Male Dancer: Bodies, Spectacles, Sexualities.* London: Routledge.

——. (2001). 'Dissolving in Pleasure: The Threat of the Queer Male Dancing Body.' In *Dancing Desires: Choreographing Sexualities on and off the Stage.* Ed. Jane Desmond. Madison: University of Wisconsin Press.

Buell, Lawrence. (1999). 'In Pursuit of Ethics.' *PMLA,* 114.1: 7–17.

Butler, Judith. (1990). 'Performative Acts and Gender Constitution: An Essay in Phenomenology and Feminist Thought.' *Performing Feminisms: Feminist Critical Theory and Theatre.* Baltimore: Johns Hopkins University Press. 270–82.

——. (1993). *Bodies That Matter: On the Discursive Limits of 'Sex'*. New York: Routledge.

——. 1997 *The Psychic Life of Power*. Palo Alto, CA: Stanford University Press.

'Cambodia Sentences 3 U.S. Men in Attack.' (2001). *Los Angeles Times*. 23 June: A8.

Cambodian Genocide Project. Available at: http://www.yale.edu/cgp/>

Cantine, Elizabeth M. Letter. (2000). *Los Angeles Times Magazine*. 16 January: 6.

Carman, Joseph. (2001). 'In Modern Dance, Male Isn't the Weaker Sex Now.' *New York Times*. 14 October: AR10+.

Carrison, Muriel Paskin. (1987). *Cambodian Folk Stories from the Gatiloke*. Rutland, VT: Tuttle.

Caruth, Cathy. (1995). 'Introduction: Trauma and Experience.' In *Trauma: Explorations in Memory*. Ed. C. Caruth. Baltimore: Johns Hopkins University Press. 3–12.

——. (1996). *Unclaimed Experience: Trauma, Narrative, and History*. Baltimore: Johns Hopkins University Press.

Charlé, Suzanne. (2001). 'With Monkeys and Giants, Rescuing a Lost World.' *New York Times*. 12 August: AR6.

Chazin-Bennahum, Judith. (1997). 'Women of Faint Heart and Steel Toes.' In *Rethinking the Sylph*. Ed. Lynn Garafola. Hanover, NH: Wesleyan University Press, 121–30.

Chen, P. (1993). *Lecture*. Trans. Toni Shapiro. Long Beach, California, 5 December.

Cheng, Meiling. (2000). 'Elia Arce's Performance Art: Transculturation, Feminism, Politicized Individualism.' *Text and Performance Quarterly*, 20.2 (April): 150–81.

Clark, Katerina and M. Holquist. (1984). *Mikhail Bakhtin*. Cambridge: Belknap/Harvard University Press.

Clifford, James. (1997). *Routes: Travel and Translation in the Late Twentieth Century*. Cambridge, MA: Harvard University Press.

Coburn, Judith. (1993). 'Dancing Back.' *Los Angeles Times Magazine*. 26 September: 14+.

Cohen, Jeffrey Jerome. (1996). 'Monster Culture (Seven Theses).' In *Monster Theory*. Ed. J. Cohen. Minneapolis: University of Minnesota Press, 3–25.

Cohen, Ted. (1978/79). 'Metaphor and the Cultivation of Intimacy.' In *On Metaphor*. Ed. Sheldon Sacks. Chicago: University of Chicago Press, 1–10.

Coleridge, Samuel Taylor. (1983). 'The Rime of the Ancient Mariners.' *The Norton Anthology of Poetry*. Ed. A. Allison et al. 3rd edn. New York: Norton, 567–81.

Connerton, Paul. (1989). *How Societies Remember*. Cambridge: Cambridge University Press.

Copjec, Joan and Michael Sorkin. (1999). *Giving Ground*. London: Verso.

Criddle, JoAn D. (1987). *To Destroy You is No Loss: The Odyssey of a Cambodian Family*. East/West Bridge Publishing.

Daly, Anne. (1995). *Done Into Dance: Isadora Duncan in America*. Bloomington: University of Indiana Press.

——. (1997). 'Classical Ballet: A Discourse of Difference.' In *Meaning in Motion*. Ed. Jane Desmond. Durham, NC: Duke University Press, 111–19.

Danto, Arthur C. (1983). *The Transfiguration of the Commonplace*. Cambridge: Harvard University Press.

Davis, Mike. (1989). 'To dance is "female."' *TDR*, 34.4 (Winter): 23–7.

——. (1992). *City of Quartz: Excavating the Future in Los Angeles*. New York: Vintage.

de Certeau, Michel. (1984). *The Practice of Everyday Life*. Vol. 1. Trans. S. Rendall. Berkeley: University of California Press.

——. (1986). *Heterologies: Discourse on the Other*. Trans. Brian Massumi. Minneapolis: University of Minnesota Press.

——. (1992a). 'Mysticism.' *Diacritics*, 22.2 (Summer): 11–25.

——. (1992b). *The Mystic Fable*. Vol. 1. Trans. Michael B. Smith. Chicago: University of Chicago Press.

——. (1997). *Culture in the Plural*. Ed. Luce Giard. Trans. Tom Conley. Minneapolis and London: University of Minnesota Press.

de Certeau, Michel, et al. (1998). *The Practice of Everyday Life Volume 2: Living and Cooking*. Ed. Luce Giard. Trans. Timothy J. Tomasik. Minneapolis: University of Minnesota Press.

Decherd, Chris. (2001). 'Making His Voice Heard In Cambodia.' *Los Angeles Times Calendar Weekend*. 26 July: 55.

DelaCerna, Sylvia. (2000). Letter. *Los Angeles Times Magazine*. 16 January: 6.

Delgado, Celeste Fraser and José Muñoz, eds. (1996). *Everynight Life: Culture and Dance in Latin/o America*. Durham, NC: Duke University Press.

Derrida, Jacques. (1995). *Archive Fever: A Freudian Impression*. Trans. E. Prenowitz. Chicago: University of Chicago Press.

——. (2001). *The Work of Mourning*. Ed. P. A. Brault and M. Naass. Chicago: University of Chicago Press.

Desmond, Jane. (1991). 'Dancing Out Difference: Cultural Imperialism and Ruth St. Denis's 'Rhada' of 1906.' *Signs* 17.1: 28–49.

——. (1997). 'Introduction.' *Meaning in Motion: New Cultural Studies of Dance*. Durham, NC: Duke University Press, 1–26.

——. ed. (2001). *Dancing Desires*. Madison: University of Wisconsin Press.

Diamond, Elin. (1996). 'Introduction.' *Performance and Cultural Politics*. Ed. E. Diamond. New York: Routledge, 1–12.

——. (1997). *Unmaking Mimesis*. London: Routledge.

Dillard, Annie. (1987). *An American Childhood*. New York: Harper & Row.

Dimock, Wai Chee. (1996). *Residues of Justice*. Berkeley: University of California Press.

Doane, Mary Ann. (1991). *Femmes Fatales*. New York: Routledge.

Dolan, Jill. (2001a). *Geographies of Learning*. Middleton, CN: Wesleyan University Press.

——. (2001b). 'Performance, Utopia, and the "Utopian Performative."' *Theatre Journal*, 53: 455–79.

——. (2005). *Utopia in Performance*. Ann Arbor: University of Michigan Press.

Dunning, Jennifer. (2001). 'In Neighborhoods Where the Ballet Is a Rare Flower.' *New York Times*. 19 August: AR22.

Doxtader, Eric. (2001). 'Loving History's Fate, Perverting the Beautiful Soul: Scenes of Felicity's Potential.' *Cultural Studies*, 15.2 (April): 206–21.

Dy, Khing Hoc. (1994). 'Khmer Literature since 1975.' In *Cambodian Culture Since 1975: Homeland and Exile*. Ed. May M. Ebihara et al. Ithaca: Cornell University Press, 27–38.

Ebihara, May M., et al., eds. (1994). *Cambodian Culture since 1975: Homeland and Exile*. Ithaca: Cornell University Press.

Edwards, Tamala H. (1998). 'No Pain, No Sweat.' *Time*. April 27: 64.

Erikson, Kai. (1995). 'Notes on Trauma and Community.' *Trauma: Explorations in Memory*. Ed. C. Caruth. Baltimore: Johns Hopkins University Press, 183–99.

Feldman, Allen. (1991). *Formations of Violence: The Narrative of the Body and Political Terror in Northern Ireland*. Chicago: University of Chicago Press.

Ferlinghetti, Lawrence. (1994). 'Underwear.' *These Are My Rivers: New And_Selected Poems, 1955–(1993).* New York: New Directions, 130.

Fisher, Jennifer. (1999). 'Pointe Shoe Confidential.' *Los Angeles Times.* Calendar Section. 5 December: 5+.

——. (2003). 'Korean troupe lifts bodies and spirits.' *Los Angeles Times.* 4 February: E4.

Fisher-Nguyen, Karen. (1994). 'Khmer Proverbs: Images and Rules.' In *Cambodian Culture since 1975: Homeland and Exile.* Ed. M. Ebihara, et al. Ithaca: Cornell University Press, 91–104.

Foster, Susan Leigh. (1992). 'Dancing Bodies.' *Zone: Incorporations.* Ed. Jonathan Crary and Sanford Kwinter. New York: Urzone, 480–95.

——. (1995). 'Choreographing History.' In *Choreographing History.* Bloomington: Indiana University Press, 2–21.

——. (1997). 'Dancing Bodies.' *Meaning In Motion: New_Cultural Studies of Dance.* Ed. Jane Desmond. Durham: Duke University Press, 235–57.

——. (1998). *Choreography and Narrative: Ballet's Staging of Story and Desire.* Bloomington: Indiana University Press.

——. (1999). 'Kinesthetic Empathies and the Politics of Compassion.' Lecture. The Getty Research Institute for the History of Art and Humanities. Los Angeles, CA. May.

Foucault, Michel. (1985). *The Use of Pleasure.* Trans. R. Hurley. Harmondsworth: Penguin.

——. (1988). 'Technologies of the Self.' In *Technologies of the Self: A Seminar with Michel Foucault.* Ed. L. Martin et al. Amherst: University of Massachusetts Press, 16–49.

Franko, Mark. (1995). *Dancing Modernism/Performing Politics.* Bloomington: University of Indiana Press.

Freud, Sigmund. (1958). 'The Uncanny.' *On Creativity and the Unconscious.* Ed. B. Nelson. New York: Harper Colophon, 122–161.

Gamble, Mandy. (1997). 'A Student Dance Ethnography: Performing the "Ideal" Aesthetic.' MA Thesis. California State University, Los Angeles.

Ganguly, Keya. (1992). 'Migrant Identities: Personal Memory and the Construction of Selfhood.' *Cultural Studies,* 6.1: 27–50.

Garafola, Lynn, ed. (1997). *Rethinking the Sylph: New Perspectives on the Romantic Ballet.* Hanover, NH: Wesleyan University Press.

Garrett, George. (1992). 'Locker Room Talk: Notes on the Social History of Football.' *Witness,* 7.2: 115–22.

Giard, Lucy. (1998). 'Times and Places.' In *The Practice of Everyday Life Volume 2: Living and Cooking.* Ed. Luce Giard. Trans. Timothy J. Tomasik. Minneapolis: University of Minnesota Press, xxxv–xlv.

Gingrich-Philbrook, Craig. (2001). 'Love's Excluded Objects: Staging Irigaray's Heteronormative Essentialism.' *Cultural Studies,* 15.2 (April): 222–8.

Goldberg, Jonathan. (1992). 'Recalling Totalities: The Mirrored Stages of Arnold Schwarzenegger.' *Differences,* 4.1 (Spring): 172–204.

Gordon, Avery. (1997). *Ghostly Matters: Haunting and the Sociological Imagination.* Minneapolis: University of Minnesota Press.

Graff, Ellen. (1997). *Stepping Left: Dance and Politics in New York City, 1928–1942.* Durham, NC: Duke University Press.

Graham, Jorie. (1980). 'The Geese.' In *Hybrids of Plants and of Ghosts.* Princeton: Princeton University Press, 38–9.

——. (1995). 'Soul Says.' *The Dream of the Unified Field: Poems 1974–(1994).* Hopewell, NJ: Ecco Press, 156.

Graham, Martha. (1998). 'I am a dancer.' In *The Routledge Dance Studies Reader.* Ed. Alexandra Carter. London: Routledge, 66–71.

Grover, J. B. (2000). Letter. *Los Angeles Times Magazine.* 16 January: 6.

Gussow, Alan. (1971). *A Sense of Place: The Artist and the American Land.* San Francisco: Friends of the Earth.

Halberstam, Judith. (1995). *Skin Shows.* Durham, NC: Duke University Press.

Hall, Kari René. (1992). *Beyond the Killing Fields.* Text by J. Getlin and K. R. Hall. Hong Kong: Asia 2000.

Hamera, Judith. (1990).'Silence that Reflects: Butoh, *Ma*, and the Crosscultural Gaze.' *Text and Performance Quarterly*, 10.1 (January): 53–60.

——. (1994). 'The Ambivalent, Knowing Male Body in the Pasadena Dance Theatre.' *Text and Performance Quarterly*, 14.3 (July): 197–209.

Hanna, Judith Lynne. (1988). *Dance, Sex and Gender.* Chicago: University of Chicago Press.

Hartman, Geoffrey H. (1993). 'Public Memory and Modern Experience.' *The Yale Journal of Criticism*, 6.2: 239–47.

Hawes, Leonard C. (1998). 'Becoming-Other-Wise: Conversational Performance and the Politics of Experience.' *Text and Performance Quarterly*, 18.4 (October): 273–99.

Hayat, Pierre. (1999). 'Preface.' *Alterity & Transcendence.* By Emmanuel Levinas. Trans. Michael B. Smith. New York: Columbia University Press, ix–xxiv.

Hayden, Dolores. (1997). *The Power of Place: Urban Landscapes as Public History.* Cambridge, MA: MIT Press.

Hearn, Jeff. (1992). *Men in the Public Eye.* London: Routledge.

Hebdige, Dick. (1988). *Hiding in the Light.* London: Comedia/Routledge.

Hesser, Amanda. (1998). 'Learning Pilates One Stretch at a Time.' *New York Times.* 3 November: D9.

Hijikata, Tatsumi. (2000a). 'Plucking off the Darkness of the Flesh.' Interview with Shibusawa Tatsuhiko. *TDR*, 44.1 (Spring): 49–55.

——. (2000b) 'To Prison.' *TDR*, 44.1 (Spring): 43–8.

Him, Chanrithy. (2000). *When Broken Glass Floats: Growing Up Under the Khmer Rouge.* New York: Norton.

Hitchcock, Peter. (1993). *Dialogics of the Oppressed.* Minneapolis: University of Minnesota Press.

Holquist, Michael. (1990). *Dialogism: Bakhtin and His World.* New York: Routledge.

'Homer vs. Patty and Selma' (1995). (2F14). *The Simpsons.* Fox. 26 February.

hooks, bell. (1994). 'Homeplace (a site of resistance).' *The Woman That I Am: The Literature and Culture of Contemporary Women of Color.* Ed. D. Soyini Madison. New York: St. Martin's, 448–54.

Horwitz, Dawn Lille. (2002). 'The New York Negro Ballet in Great Britain.' In *Dancing Many Drums.* Ed. T. F. DeFrantz. Madison: University of Wisconsin Press, 317–39.

How Nice to See You Alive. (1989). Dir. Lucia Murat.

Hutcheon, Linda. (1988). *A Poetics of Postmodernism: History, Theory, Fiction.* New York: Routledge.

Irigary, Luce. (1996). *I Love to You: Sketch of a Possible Felicity in History.* Trans. Alison Martin. New York: Routledge.

Jackson, John Brinckerhoff. (1984). *Discovering the Vernacular Landscape*. New Haven: Yale University Press.

Jackson, Shannon. (2000). *Lines of Activity: Performance, Historiography, Hull-House Domesticity*. Ann Arbor: University of Michigan Press.

Jones, Amelia. (1998). *Body Art: Performing the Subject*. Minneapolis: University of Minnesota Press.

Kiernan, Ben. (1996). *The Pol Pot Regime: Race, Power, and Genocide in Cambodia Under the Khmer Rouge, 1975–79*. New Haven: Yale University Press.

Kirkland, Gelsey. (1986). *Dancing on My Grave*. New York: Berkley/Doubleday.

Kisselgoff, Anna. (2002). 'In the Ranks of Idols.' *New York Times*. 13 June: B1+.

Klein, Norman M. (1997). *The History of Forgetting*. London: Verso.

Klein, Susan Blakeley. (1988). *Ankoku Buto: The Premodern and Postmodern Influences on the Dance of Utter Darkness*. Ithaca: Cornell University East Asia Program.

Kofman, Eleonore and E. Lebas. (1996). 'Lost in Transposition – Time, Space and the City.' *Henri Lefebvre: Writings on Cities*. Ed. and Trans. Eleonore Kofman and Elizabeth Lebas. Oxford: Blackwell, 3–60.

Koritz, Amy. (1995). *Gendering Bodies/Performing Art: Dance and Literature in Early Twentieth Century British Culture*. Ann Arbor: University of Michican Press.

Kuniyoshi, Kazuko. (1986). 'Butoh Chronology: 1959–(1984).' *TDR*, 30.2 (Summer): 127–41.

Kurihara, Nanako. (2000). 'Hijikata Tatsumi: The Words of Butoh.' *TDR*, 44.1 (Spring): 12–28.

Kuspit, Donald. (1986). 'Concerning the Spiritual in Contemporary Art.' In *The Spiritual in Art: Abstract Painting 1890–(1985)*. New York: Abbeville Press, 313–25.

LaFreniere, Bree. (2000). *Music Through the Dark*. Honolulu: University of Hawaii Press.

Langellier, Kristin. (1999). 'Personal Narrative, Performance, Performativity: Two or Three Things I Know For Sure.' *Text and Performance Quarterly*, 19.2 (April): 125–44.

Lebeau, Vicky. (1991). '"You're My Friend": *River's Edge* and Social Spectatorship.' *Camera Obscura*, 25/26 (January/May): 251–72.

Lee, Carol. (2002). *Ballet in Western Culture: A History of its Origins and Evolution*. New York: Routledge.

Lefebvre, Henri. (1996). *Writings on Cities*. Trans. and Ed. E. Kofman and E. Lebas. London: Blackwell.

Leivick, Laura. (2000). 'Where Boys Can Learn to Shape Up.' *New York Times*. 28 May: AR6+.

Lionnet, Françoise. (1995). *Postcolonial Representations*. Ithaca: Cornell University Press.

Looseleaf, Victoria. (2002). 'Oguri's Butoh Style Inspired by Getty's Setting.' *Los Angeles Times*. 10 June: F11.

Lyotard, Jean François. (1988). *The Inhuman*. Trans. G. Bennington and R. Bowlby. Stanford: Stanford University Press.

——. (1997). *Postmodern Fables*. Trans. G. van den Abbeele. Minneapolis: University of Minnesota Press.

Madison, D. Soyini, ed. (1994). *The Woman That I Am: The Literature and Culture of Contemporary Women of Color*. New York: St. Martin's.

——. (1999). 'Performing Theory/Embodied Writing.' *Text and Performance Quarterly*, 19.2 (April): 107–24.

——. (2000). 'Oedipus Rex at *Eve's Bayou* Or The Little Black Girl Who Left Sigmund Freud in the Swamp.' *Cultural Studies*, 14.2 (April): 311–40.

Mamet, David. (1989). *Some Freaks*. New York: Viking.

Martin, Randy. (1998). *Critical Moves*. Durham, NC: Duke University Press.

Matthews, Pamela R. and David McWhirter, eds. (2003). 'Introduction: Exile's Return? Aesthetics Now.' *Aesthetic Subjects*. Minneapolis: University of Minnesota Press, xiii–xxviii.

Mayol, Pierre. (1998a). 'The Neighborhood.' In *The Practice of Everyday Life. Volume 2: Living and Cooking*. Minneapolis: University of Minnesota Press, 35–70.

——. (1998b). 'Propriety.' In *The Practices of Everyday Life. Volume 2: Living and Cooking*. Trans. Timothy J. Tomasik. Minneapolis: University of Minnesota Press, 15–34.

——. (1998c). 'The Street Trade.' In *The Practice of Everyday Life. Volume 2: Living and Cooking*. Minneapolis: University of Minnesota Press, 71–83.

McRobbie, Angela. (1997). 'Dance Narratives and Fantasies of Achievement.' In *Meaning in Motion: New Cultural Studies of Dance*. Ed. Jane C. Desmond. Durham, NC: Duke University Press, 207–31.

Mitchell, W. J. T. (1998). *The Last Dinosaur Book*. Chicago: University of Chicago Press.

Modleski, Tania. (1991). *Feminism Without Women*. New York: Routledge.

'Morning Report: "Billy Elliot" Inspires Young Male Dancers.' (2002). *Los Angeles Times*. 16 April: F2.

Morris, David B. (1991). *The Culture of Pain*. Berkeley: University of California Press.

Morris, Gay, ed. (1996). *Moving Worlds*. New York: Routledge.

Morris, Meaghan. (1988). 'Banality in Cultural Studies.' *Discourse*, 10.2: 3–29.

Morse, Susan. (2001). 'Recent Court Ruling Puts Pilates Name Up for Grabs.' *Los Angeles Times*. January 15: S8.

Morson, Gary Saul. (1994). *Narrative and Freedom: The Shadows of Time*. New Haven: Yale University Press.

Mulvey, Laura. (1975). 'Visual Pleasure and Narrative Cinema.' *Screen*, 16.3: 6–18.

Mydans, Seth. (1993). 'Khmer Dancers Try To Save an Art Form Ravaged by War.' *New York Times*. 30 December: C11+.

——. (1999). 'Smiles Were Rare: Khmer Rouge Photography.' *New York Times*. 24 January: WK5.

——. (2000). 'Khmer Dance in a Lesson for Khmer Rouge.' *New York Times*. 2 May: B1.

——. (2001). 'For Cambodian, a Life Transformed by Dance.' *New York Times*. 9 March: A4.

Myerhoff, Barbara. (1979). *Number Our Days*. New York: Touchstone.

——. (1992). *Remembered Lives: The Work of Ritual, Storytelling, and Growing Older*. Ed. M. Kaminsky. Ann Arbor: University of Michigan Press.

Naficy, Hamid. (1993). *The Making of Exile Cultures: Iranian Television in Los Angeles*. Minneapolis: University of Minnesota Press.

Nagourney, Eric. (1999). 'Vital Signs: When on Their Toes Means at Some Risk.' *New York Times*. 7 December: D8.

——. (2000). 'Vital Signs: Hidden Hazards of Tutus and Toe Shoes.' *New York Times*. 24 October: D8.

Nath, Vann. (1998). *A Cambodian Prison Portrait: One Year in the Khmer Rouge's S – 21*. Trans. M. C. Nariddh. Bangkok: White Lotus Press.

Ness, Sally Ann. (1996). 'Observing the Evidence Fail: Difference Arising from Objectification in Cross-Cultural Studies of Dance.' In *Moving Words*. Ed. G. Morris. New York: Routledge, 245–69.

Ngor Haing. (1987). *A Cambodian Odyssey*. New York: Warner Books.

Nunez, Sigrid. (1995). *A Feather on the Breath of God*. New York: HarperCollins.

Oliver, Kelly. (2001). *Witnessing*. Minneapolis: University of Minnesota Press.

Pape, Eric. (2001). 'The Remote Control Revolution.' *Los Angeles Times Magazine*. 24 June: 14+.

Palmer, David L. (1998). 'Virtuosity as Rhetoric: Agency and Transformation in Paganini's Mastery of the Violin.' *Quarterly Journal of Speech*, 84.3 (August): 341–57.

Patraka, Vivian. (1999). *Spectacular Suffering: Theatre, Fascism, and the Holocaust*. Bloomington: Indiana University Press.

Pearlstone, Zena. (1990). *Ethnic L.A.* Beverly Hills: Hillcrest Press.

Peters, John Durham. (1999). *Speaking Into the Air: A History of the Idea of Communication*. Chicago: University of Chicago Press.

Peterson, Michael. (1997). *Straight White Male*. Jackson: University of Mississippi Press.

Phelan, Peggy. (1995). 'Thirteen Ways of Looking at *Choreographing Writing*.' In *Choreographing History*. Ed. Susan Leigh Foster. Bloomington: Indiana University Press, 200–10.

——. (1998). 'Introduction: The Ends of Performance.' In *The Ends of Performance*. Ed. P. Phelan and J. Lane. New York: New York University Press, 1–19.

Phim, Toni Samantha and Ashley Thompson. (1999). *Dance in Cambodia*. Oxford: Oxford University Press.

Pich Tum Kravel. (1990). Program Notes, 'Classical Dance Company of Cambodia.' Los Angeles Festival: 3–4.

Pollock, Della. (1998a) 'Introduction: Making History Go.' In *Exceptional Spaces: Essays in Performance and History*. Ed. D. Pollock, Chapel Hill: University of North Carolina Press. 1–45.

——. (1998b). 'Performing Writing.' In *The Ends of Performance*. Ed. P. Phelan and J. Lane. New York: New York University Press, 73–103.

——. (1999). *Telling Bodies Performing Birth*. New York: Columbia University Press.

——. (2000). 'Finding Pleasure in Margaret.' Presentation. National Communication Association Annual Conference. Seattle, WA.

——. (2001). 'Editor's Note on *Performing Love*.' *Cultural Studies*, 15.2 (April): 203–5.

Pran, Dith. (1997). *Children of Cambodia's Killing Fields: Memoirs by Survivors*. Ed. K. DePaul. New Haven: Yale University Press.

Pudelek, Janina. (1997) 'Ballet Dancers at Warsaw's Wielkie Theater.' In *Rethinking the Sylph*. Ed. Lynn Garafola. Hanover, NH: Wesleyan University Press, 143–63.

Rauzi, Robin. (1998). 'Taking Aim at Inspiration.' *Los Angeles Times*. 17 May: Calendar 56+.

Sachs, Muriel. (2000). Letter. *Los Angeles Times Magazine*. 16 January: 6.

Sadono, Regina Fletcher. (1999). 'Performing Symptoms.' *Text and Performance Quarterly*, 19.2 (April): 159–71.

Sam, Sam-Ang. (1994). 'Khmer Traditional Music Today.' In *Cambodian Culture since 1975: Homeland and Exile*. Ed. M. Ebihara etal. Ithaca: Cornell University Press, 39–47.

Sassen, Saskia. (2000). *Cities in a World Economy*. 2nd edn. Thousand Oaks, CA: Pine Forge Press.

Sayers, Lesley-Anne. (1997). 'Madame Smudge, Some Fossils, and Other Missing Links: Unearthing the Ballet Class.' *Dance In The City*. Ed. Helen Thomas. New York: St. Martin's, 130–48.

Scarry, Elaine. (1985). *The Body in Pain: The Making and Unmaking of the World*. New York: Oxford University Press.

——. (1999). *On Beauty and Being Just*. Princeton: Princeton University Press.

Schechner, Richard. (1985). *Between Theatre and Anthropology*. Philadelphia: University of Pennsylvania Press.

Schneider, Rebecca. (1997). *The Explicit Body in Performance*. New York: Routledge.

Scholes, Robert. (1989). *Protocols of Reading*. New Haven: Yale University Press.

——. (1992). 'Response' to 'Reading Robert Scholes: A Symposium.' *Text and Performance Quarterly*, 12.1 (January): 75–8.

Sedgwick, Eve Kosofsky. (1985). *Between Men: English Literature and Male Homosocial Desire*. New York: Columbia University Press.

——. (1990). *Epistemology of the Closet*. Berkeley: University of California Press.

Segal, Lewis. (1998). 'Oguri Still Challenging Physical Limits.' *Los Angeles Times*. 25 May: F3.

——. (2004). 'Late for the Dance.' *Los Angeles Times*. 8 August: E4.

Sells, Michael A. (1994). *Mystical Languages of Unsaying*. Chicago: University of Chicago Press.

Shattuck, Kathryn. (2002). 'Teaching Hairy Guys in Tutus to Take Flight.' *New York Times*. 11 August: AR6+

——. (2003). 'To Dance, Perchance to Be Seen.' *New York Times*. 22 June: AR27.

Sippl, Diane. (1994a.). '...Even As Also I Am Known: Vicarious Miscegenation on Postcolonial Screens.' *CineAction*, no. 3 (February): 23–42.

——. (1994b). 'Terrorist Acts and "Legitimate" Torture in Brazil: *How Nice to See You Alive.*' *Discourse*, 17.1 (Fall): 77–92.

Smith, F. (1989). *Interpretive Accounts of the Khmer Rouge Years: Personal Experience in Cambodian Peasant World View*. Madison: Center for Southeast Asian Studies, University of Wisconsin.

Soja, Edward W. (1996). *Thirdspace*. London: Blackwell.

Stallybrass, Peter. (1999). 'Worn Worlds: Clothes, Mourning and the Life of Things.' In *Cultural Memory and the Construction of Identity*. Ed. D. Ben-Amos and L. Weissberg. Detroit: Wayne State University Press.

Stein, Bonnie Sue. (1986). 'Butoh: "Twenty Years Ago We Were Crazy, Dirty, and Mad."' *TDR*, 30.2 (Summer): 107–25.

Steinberg, Morleigh, dir. (1994). *Traveling Light*. Video.

Stevens, Wallace. (1982). 'The Idea of Order at Key West.' *The Collected Poems*. New York: Vintage, 128–30.

Stewart, Jocelyn Y. (1997). 'Reaching Out to Cambodians.' *Los Angeles Times*. 17 November: B1.

——. (2001). 'One Step at a Time, Dream of Ballet Company Takes Shape.' *Los Angeles Times*. 25 August: B1+.

Stewart, Susan. (1993). *On Longing: Narratives of the Miniature, the Gigantic, the Souvenir, the Collection*. Durham, NC: Duke University Press.

'Sunday Q&A: Plié, Pirouette, Ouch!' (2001). *New York Times*. 9 September: 25.

Szymusiak, Molyda. (1999). *The Stones Cry Out: A Cambodian Childhood, 1975–1980*. Trans. L. Coverdale. Bloomington: Indiana University Press.

Tanaka, Min. (1986). 'Farmer/Dancer or Dancer/Farmer.' Interview with Bonnie Sue Stein. *TDR*, 30.2 (Summer): 142–51.

Taylor, Diana. (2003). *The Archive and the Repertoire: Performing Cultural Memory in the Americas*. Durham, NC: Duke University Press.

Terdiman, Richard. (1991). 'The Response of the Other.' *Diacritics*, 22.2 (Summer): 2–10.

The Turning Point. (1977). Dir. Herbert Ross.

Thomas, Helen. (1996). 'Do You Want to Join the Dance? Postmodernism/Poststructuralism, the Body, and Dance.' In *Moving Worlds: Re-writing Dance*. Ed. Gay Morris. New York: Routledge, 63–87.

Thrussell, Paul. (2001). '"Billy" & me.' *Boston Sunday Globe*, 27 May: M6+.

Tsing, Anna Lowenhaupt. (2005). *Friction: An Ethnography of Global Connection*. Princeton: Princeton University Press.

Tuan, Yi-Fu. (1977). *Space and Place: The Perspective of Experience*. Minneapolis: University of Minnesota Press.

———. (1992). 'Place and Culture: Analeptic for Individuality and the World's Indifference.' In *Mapping American Culture*. Ed. W. Franklin and M. Steiner. Iowa City: University of Iowa Press, 27–49.

Turnbull, Robert. (1999). 'Reconstructing Khmer Classics From Zero.' *New York Times*. 25 July: AR6.

———. (2001). 'Making Dance of the Killing Fields.' *New York Times*. 24 June: AR6.

Ung, Loung. (2000). *First They Killed My Father: A Daughter of Cambodia Remembers*. New York: HarperCollins.

Vendler, Helen. (1986). *Wallace Stevens: Words Chosen Out of Desire*. Cambridge: Harvard University Press.

Villa, Raúl Homero and George J. Sánchez. (2004). 'Los Angeles Studies and the Future of Urban Cultures.' *American Quarterly*, 56.3 (September): 499–505.

Weiss, Allen S. (2004). 'Ten Theses on Monsters and Monstrosities.' *TDR*, 48.1 (Spring): 124–5.

Welaratna, Usha. (1993). *Beyond the Killing Fields: Voices of Nine Cambodian Survivors In America*. Stanford: Stanford University Press.

Welty, Eudora. (1978). 'Place in Fiction.' *The Eye of the Story*. New York: Vintage.

'What Dance Has to Say About Beauty.' (2000). *New York Times*. 23 July: AR26+.

White, Hayden. (1995). 'Bodies and Their Plots.' In *Choreographing History*. Ed. Susan Leigh Foster. Bloomington: Indiana University Press, 230–34.

Wolff, Janet. (1997). 'Reinstating Corporeality: Feminism and Body Politics.' In *Meaning in Motion*. Ed. Jane C. Desmond. Durham, NC: Duke University Press, 81–99.

Wride, Nancy. (2001). 'Cambodian Community Makes Banner Statement.' *Los Angeles Times*. 15 July: B1.

Wyschogrod, Edith. (1985). *Spirit in Ashes: Hegel, Heidegger, and Man-Made Mass Death*. New Haven: Yale University Press.

Yaeger, Patricia. (1996). 'Introduction.' *The Geography of Identity*. Ed. P. Yaeger. Ann Arbor: University of Michigan Press, 1–38.

Zarrilli, Philip B. (2002). 'The Metaphysical Studio.' *TDR*, 46.2 (Summer): 157–70.

Index

Abu-Lughod, Janet L., 9, 11
Adato, Allison, 78–8, 83, 87, 91
Aesthetics, 3–5, 210–12
And technique, 5
Ahearne, Jeremy, 22, 202
American Ballet Theatre (ABT), 79, 80, 84 141
Amin, Ash, and Nigel Thrift, 1, 2, 15, 17, 23, 58, 210, 213 n. 8
Anawalt, Sasha, 76, 80
Anderson, Benedict, 147
Answerability, 146–70, 221 n. 9
Ambivalence of, 149
Apsara as technology of, 153–6
As palimpsest, 149–59
Definition, 149
Unitary aspects of, 149–52
Apophasis, 193
Apsara, 154–9, 165–7
As 'epic' body, 155–9
Material circumstances of, 166–7
Aristotle, 1
Attridge, David, 188, 191–2
Auslander, Philip, 46

Babitz, Eve, 17, 27
Bachelard, Gaston, 70–1, 76
Bakhtin, Mikhail, 72–5, 149–52, 155
Answerability, 146–70, 221 n. 9
Chronotope, 72–91, 93–110
Balanchine, George, 78, 108, 113, 122
Ballet:
And space, 65–72
And time, 64–5
As vernacular landscape, 60–1
French in, 68–9
Gender in, 102–7, 120–6, 218 n. 18, 218 n. 2
Heteronormativity in, 103, 107, 134
Nostalgia in, 112–15
Organization of body, 31, 57, 66–7

Pain in (*see also* Pointe shoes), 99–106, 127–8
Pointe shoes, 100–3, 217 n. 14 and 15
Popular culture representations of, 65, 74–7, 118–20
Weight in, 217 n. 13
Banes, Sally, 75, 122, 175, 189, 220 n. 2
Barkin, Boaz, 44
Barthes, Roland, 6, 32–3, 38, 45, 51
Bataille, Georges, 25, 41, 54
Battaglia, Debbora, 114
Beauty, 106–7, 137, 203–5, 221 n. 13
Berlant, Lauren, and Michael Warner, 15, 17, 18, 209, 213 n. 1
Bennett, Jane, 12–13, 172, 204
Bennett, Susan, 2
Bettleheim, Bruno, 171
Bhabha, Homi, 98–200
Bishop, Elizabeth, 195
Blau, Herbert, 50
Body Weather Laboratory, 43–4
Philosophy of, 43
Booth, Wayne, 209
Bourdieu, Pierre, 89, 216 n. 12
Brandon, James, 146
Brecht, Bertold, 207
Briginshaw, Valerie, 10, 25, 122
Brodsky, Joseph, 150
Brody, Liz, 29
Brookings Institution, 9
Brown, Lyn Mikel, and Carol Gilligan, 97–8
Browning, Barbara, 29, 177, 214 n. 7, 220 n. 2
Brumberg, Joan Jacobs, 96–7, 104
Burns, Roger, 47, 50, 51, 196, 201
Burris, Jamie, 44, 50, 51
Burt, Ramsay, 53, 108, 116, 120–3, 218 n. 19
Bustillo, Gilma, 61, 62, 98